Air Pollution and Acid Rain:
The Biological Impact

Alan **Wellburn**

DIRECTOR OF BIOCHEMISTRY,
DEPARTMENT OF BIOLOGICAL SCIENCES,
UNIVERSITY OF LANCASTER, UK

AIR POLLUTION AND ACID RAIN:
The Biological Impact

Longman
Scientific &
Technical

Copublished in the United States with
John Wiley & Sons, Inc., New York

Longman Scientific & Technical,
Longman Group UK Limited,
Longman House, Burnt Mill, Harlow,
Essex CM20 2JE, England
and Associated Companies throughout the world.

Copublished in the United States with
John Wiley & Sons, Inc., 605 Third Avenue, New York, NY 10158

First published 1988

British Library Cataloguing in Publication Data
Wellburn, Alan
   Air pollution and acid rain: the
   biological impact.
   1. Air – Pollution – Physiological
   effect
   I. Title   II. Series
   574.5'222   QH545.A3

ISBN 0-582-01464-6

Library of Congress Cataloguing in Publication Data
Wellburn, Alan, 1940–
   Air pollution and acid rain.

   Includes bibliographies and indexes.
   1. Air pollution – Environmental aspects. 2. Acid
rain – Environmental aspects. I. Title. II. Series.
QH545.A3W45 1988   574.5'222   87–16742
ISBN 0–470–20887–2 (USA only)

Set in Linotron 202 10/11 pt Ehrhardt Roman

Produced by Longman Singapore Publishers (Pte) Ltd.
Printed in Singapore

Preface

'This coale . . . . . flies abroad . . . . . .
and in the Spring-time besoots all the Leaves,
so as there is nothing free from its universal contamination. . .
. . . . . and kills our Bees and Flowers abroad,
suffering nothing in our Gardens to bud,
    display themselves, or ripen.'

*Report entitled 'Fumifugium, or the inconvenience of the Air and Smoke of
London dissipated' written for King Charles II in 1661 by John Evelyn
(1620–1706)*

Pollution has always been an emotive topic irrespective of whether
it occurs in the air, on land or at sea. In the latter half of this century
much attention has been drawn to air pollution caused by modern
methods of communication although some cradles of the Industrial
Revolution must have experienced local levels of high pollution quite
unimaginable by present standards and we must never forget that such
progress upon which our society is built was bought at great cost to
the health and the immediate surroundings of those involved.

Campaigns against the more widespread effects of atmospheric
pollution are constantly mounted by the media, of which the printed
word is only a part; meanwhile scientific knowledge still advances on
all fronts. This greater appreciation allows a more fundamental
understanding of the consequences of pollution upon the whole

environment and all those who live in it, which includes the polluters, the polluted, the media and the researchers – hence the emotion.

If one turns to books for information on the subject the problem is not the lack of it but the abundance. On the topics of air pollution and acid rain, library shelves in centres of learning are full of scientific journals and perhaps a hundred books of expert opinion on one or more of the various aspects involved. The difficulty is making out what it all means and what the consequences are to all of us and to the world in which we live. One could turn to appreciations by journalists who thrive on sensationalism to make their points, valid though some of them may be. This may be suitable for socio – politico – economic appreciations which avoid the scientific fundamentals. The book on *Acid Rain* by Elsworth (Pluto Press, 1984) is a good example of this approach. If on the other hand a scientific understanding of the background mechanisms, without the need to become an expert in each area, is desired then there are very few sources to rely upon. Very often texts tend to concentrate on one aspect of air pollution (i.e. air chemistry or the effects or the consequences for vegetation or animals) without linking them together. This book attempts to do this and at the same time show how this term 'acid rain' is a *part* of atmospheric pollution, how the complicated chemistry of the atmosphere is a series of interrelated reactions, and how the components involved interact with others, with vegetation, with microbes and with ourselves at the molecular level.

In studies of the consequences of air pollution there is a general tendency to treat the living systems which they affect as though they were discrete and separate rather than interreactive. A great deal of effort goes into quantifying how much and what type of pollution exists in the atmosphere, by what means it moves around and in describing the outward changes of living systems: how much they change colour or show leaf defects if they are plants, what types of lung ailments they have if they are humans, or if as fish they disappear from lakes. One then is left with an immense catalogue of ills and ailments which gets longer as time passes. Indeed, if one considers the real world where atmospheric pollutants nearly always exist as mixtures rather than as single chemicals, the list of possible consequences upon different biological systems expands alarmingly. The only sensible course of action is to delve into the basic mechanisms by which each of the different pollutants enters and attacks living systems, to examine how they in turn resist change and, if they fail, to assess the nature and consequences of metabolic changes within all living systems. To do this, even as an expert, one needs to be a combination of a physicist, chemist, biochemist, physiologist, microbiologist, animal toxicologist and plant pathologist, a difficult task given the multitude of information that exists. Nevertheless, there are a number of broad principles that govern the effects of atmospheric pollutants upon living systems which can be gathered together and that is what this book is all about. It attempts to examine these mechanisms in plants and animals in rather more detail than other books on air pollution. Ultimately, this is the only way to make sense of the mass of information concerning the visible and outward changes upon the variety of living systems which has been ascribed to atmospheric pollutants – especially if we are ever to progress to an understanding of how combinations of pollutants interact with each other in the real world. It is therefore unavoidable that a prior understanding of basic chemistry is necessary in order to appreciate these mechanisms but any extra biochemistry that is needed has been included either in the text or in the Appendix to assist the understanding of the reader.

In order to consider internal changes in detail, less space is devoted to describing individual outward effects of single air pollutants on different species. For example, it would have been possible to show in colour the large numbers of different visible defects on leaves which have been ascribed to individual air pollutants. Unfortunately, every species is different and, as will be demonstrated, there are many other environmental and pathogenic agents (not air pollutants) which have the potential to cause similar visible injuries so that unless one is a highly practised specialist one would not derive much practical benefit in the field because the danger of misinterpretation is so great. In the case of human injury, one would naturally leave the diagnosis of bronchitis to clinicians and the attribution by experts of individual external changes to vegetation caused by air pollution is no different.

The major harm caused by air pollutants and acidic precipitation, locally and globally, is ultimately shown as a lack of growth of plants or poorer health of animals. This may mean many different things have happened but usually it occurs because of difficulties with the regulation of normal events that go on in living cells. It has been said that 'the cell is a self-centered comfort machine' which might be an overstatement but the principle of regulatory events acting in concert together to maintain a balance of cellular activities within pre-set limits has been well known to physiologists for a long time and was given the name 'homeostasis' by Cannon, or 'the wisdom of the body' as he called these automatic controls.

As this book will show, homeostasis has a cost and that extra regulation or repair as a consequence of air pollution diverts more energy away from growth, reproduction and the other necessities of life. This means that loss of growth potential by crops or forests, or less fish fry developing in acidic waters, or poorer health of people living in polluted areas are the most important consequences of atmospheric pollution. Furthermore, any diversion of energy away from growth means that any detrimental influence upon the means of energy production magnifies the problem. As will be shown, this is precisely what some air pollutants do. They attack the powerhouses of cells or the regulatory mechanisms which control them.

Early chapters are broadly concerned with different pollutants or groups of pollutants and most of them have three parts. Sources and the atmospheric chemistry of specific pollutants are covered in the first section of each chapter, the second usually deals with plants and the third is concerned with animals or human health. In some instances, groups of pollutants are considered together in order to emphasise any differences in response as well as any similarities between them. Additional sections or changes from this pattern are included as appropriate. For example, an examination of fish physiology in the presence of freshwater acidity occurs in Ch. 4 which is concerned with acid rain. The book culminates with two diagrams (Fig. 7.6 and 7.8) which represent an overview and recapitulation of the various themes of the book. The intention is to lead the reader towards an understanding of the various processes involved in these diagrams but there is no reason why one may not have a look at these parts of the ending from time to time if that is your inclination.

Rather than disrupt the text with references, a selected bibliography has been included at the end of each chapter whilst source details for some of the figures and tables are included in the appropriate selected bibliography. Otherwise they are identified in the table footnotes or figure legends. Most of the points raised in the text can be followed up in detail from these sources by anyone who wishes to do so. To those experts who recognise their own observations being mentioned without direct reference to them or who recognise defects or omissions, I apologise but trust they appreciate that the major intention is to convey a broad understanding to others. Due acknowledgement has been made in the appropriate legends to all those who allowed me to reproduce some of the figures from their work and my grateful thanks is extended to Mrs Barbara Smith for her patient work on the word processor, to my wife Florence for carefully checking manuscripts and proofs, and to my son Richard for all his efforts towards the preparation of the index.

Alan Wellburn
Lancaster 1987

Contents

'I do not mean to say that all rain is acid;
it is found with so much ammonia in it as to overcome the
acidity;
but in general, I think, the acid prevails in the town'

From 'Air and Rain – The Beginning of a Chemical Climatology
written in 1872 by Robert Angus Smith.

Acknowledgements

We are indebted to the following for permission to reproduce copyright material:

Bayerishe Staatsministerium für Ernährung, Landwirtschaft und Forsten for fig. 7.2 from p. 32 of *Wald in Gefahr*; Der Bundesminister für Forschung und Technologie for fig. 1.1 from fig. 6, p. 11 of 'Initial Report on Research on Environmental Damage to Forests' (CCRAC 84–17, Annex 1, 18 Jan 1984); CRC Press Inc. and the authors for fig. 7.6 from fig. 1 by P. H. Freer-Smith and A. Wellburn in 'Models in Plant Physiology/Biochemistry' edited by Newman and Wilson; the author, E. B. Ford for figs. 7.4 a & b from plates 14 & 15 in *Ecological Genetics*; the authors, I. Nicholson *et al* for fig. 7.1 from fig. 3, p. 145 of 'Ecological Impact of Acid Precipitation', 1980 SNSF Project, Oslo; D. Reidel Publishing Company and the authors for fig. 7.3 from fig. 3 by H. Flühler, p. 309 of 'Effects of accumulation of air pollutants in forest ecosystems' edited by Ulrich and Pankrath, 1983 and figs. 2.1 & 3.8 from figs. 3 & 4 by R. A. Cox S. A. Penkett in 'Acid Deposition' edited by Beilke & Elshout; the authors, D. Sutcliffe and T. Carrick for fig. 4.5 from fig. 1 of 'Effects of acid rain on waterbodies in Cumbria' in *Pollution in Cumbria*, ITE Symposium May 1985; The Watt Committee on Energy Ltd. and F. B. Smith for fig. 4.1 from fig. 1.1 of *Watt Committee Report* No. 14, August 1984; WHO Centre, Canada and the authors for fig. 4.2 from fig. 1 of *The Water Quality Bulletin*, April 1983 by D. Whelpdale derived from a diagram in 'Effects of Acid Precipitation on Terrestrial Ecosystems' by G. Gravenhorst.

To Victoria

1

Introduction

'We have first raised a dust
and then complain we cannot see'.

From the introduction to the Principles of Human Knowledge by Bishop Berkeley
(1685–1753).

1.1 DEFINITIONS AND TERMS

A chemical that is in the wrong place at the wrong concentration may
be either called a 'pollutant' or a 'contaminant'. The distinction
resides in the ability to show that a pollutant is capable of causing
(or has the potential to cause) damage to man and the environment.
The study of atmospheric or air pollution covers all those pollutants
that are emitted into the atmosphere from the land and oceans (usually
as gases or particulates) which then directly or indirectly degrade the
physical and biological systems on the surface of the Earth. In passing
through the atmosphere they may move from a dry gaseous phase into
a liquid phase before returning to Earth; one aspect of which has
been given the title 'acid rain'. A less-emotive and more general
descriptive word to cover both non-acidic and acidic precipitation by
means of water droplets or snowflakes is 'wet deposition' – a phrase
much lacking in journalistic merit. Nevertheless the terms 'wet' and
'dry deposition' encompass wider concepts and distinguish the two
main pathways by which atmospheric pollutants are returned to the

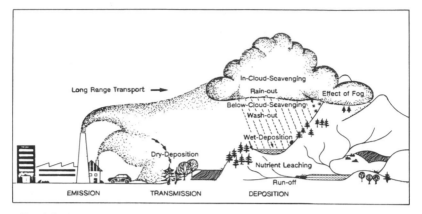

Fig. 1.1   Emission transport and mechanisms of deposition of atmospheric pollution. (Courtesy of the Bundesminister fur Forschung und Technologie, BRD.)

surface of the Earth. Other technical terms like 'rain-out', 'run-off' and even 'wash-out' are also frequently used. Figure 1.1 attempts to gather these various processes together, explain them and show the two major routes of deposition.

Emissions of pollutants into the atmosphere are induced by the activities of man but natural emissions of similar substances by volcanoes and swamps have always occurred long before the appearance of man and have always caused considerable disruption to the local environment. Such natural emissions and sources must always be taken into account along with those caused by man. It is often convenient to think of air pollutants being emitted from 'sources' and being removed from the atmosphere by 'sinks'. The term 'anthropogenic emissions' has sometimes been used to separate the man-made from natural emissions. But 'anthropogenesis' means the study of the origin of man so 'anthropogenic' is clearly a wrong use of the word. 'Man-made' or 'man-induced' are ungainly phrases but more accurate.

Damage by air pollutants caused directly by gases, or by corrosion by air pollutants in liquid form, to metals, fabrics and materials used by man is clearly apparent but the biological effects of air pollutants directly upon man or upon the living systems that surround him are far more significant. Frequently, biological effects are detected much faster at lower concentrations than those upon non-living materials. Although damage to inanimate materials has a huge economic cost it was generally accepted in the past that danger to human health was the reason for control of general pollution. More recently, especially in regard to air pollution, controls have been imposed to protect vegetation and biological ecosystems upon which man depends. This is because in the case of nearly every air pollutant, except carbon monoxide and certain hydrocarbons, the plants that make up major parts of these ecosystems are more sensitive to the presence of air pollutants than man. In each of the chapters that follow, the pathway of individual pollutants is considered from the point of emisson into the atmosphere and from there into either plants or animals. Once inside living things the effects of the products of pollution within these two systems are examined.

Air pollutants exist in gaseous or particulate forms. The latter are particularly diverse in character, consisting of many different components (see Sect. 1.7) whilst gaseous air pollutants may be separated into primary and secondary forms. Primary air pollutants like sulphur dioxide, most oxides of nitrogen, carbon monoxide and unburnt hydrocarbons are those directly emitted into the atmosphere. Secondary air pollutants like ozone and photo-activated irritants such as peroxyacetyl nitrite are formed as a consequence of subsequent reactions in the atmosphere involving primary air pollutants and other agents such as strong light.

1.2 UNITS AND CONCENTRATIONS

Concentrations of gaseous air pollutants are expressed either as mass per unit volume (i.e. micrograms per cubic metre or $\mu g \, m^{-3}$) or as volume per unit volume (which is also known as the 'volume mixing ratio'). This is expressed either as parts per million (ppm or, better, ppmv) or parts per billion (10^{-9}). Neither $\mu g \, m^{-3}$ nor ppmv, etc., are satisfactory. It would be much better if mass per unit mass (i.e. $\mu g \, g^{-1}$ or $mg \, kg^{-1}$) were to be used which would then bring air concentrations into line with standard SI measurements of substances in solids or liquids. In these latter cases $1 \, \mu g \, g^{-1}$ or $1 \, mg \, kg^{-1}$ are the same as 1 ppm or 0.0001 %. Unfortunately it is difficult immediately to conceive of what volume of air corresponds to 1 g or 1 kg and this varies with altitude anyway. Furthermore, any mass per unit volume measurement must take into account changing temperatures as well as pressure whereas the volume mixing ratio of an ideal gas is independent of temperature or pressure. Indeed, most air pollutants behave for all practical purposes as ideal gases. The relationships between the two are given by the following expressions:

$$\mu g \, m^{-3} = \frac{ppbv \times M \times 10^{-3}}{V_0} \times \frac{T_0}{T} \times \frac{P}{P_0}$$

$$ppbv = \frac{\mu g \, m^{-3} \times V_0 \times 10^{-3}}{M} \times \frac{T}{T_0} \times \frac{P_0}{P}$$

where V_0 is the molar volume of an ideal gas which is equivalent to 22.4×10^{-3} m^3 mol^{-1} when standard temperature (273° Kelvin or K) and pressure (1 atm or 101.3 kPa) prevail. For any other condition, appropriate corrections for volume changes due to differences in temperature and pressure must be made. This is laborious and must be done each time. A simple computer program in BASIC which will adapt to any type of microcomputer to do this is to be found in the Appendix (Sect. 8.1). Approximate interconversions are also provided in Tables 8.1 and 8.2.

The major point to be drawn from these formulae is that a mass per unit volume expression does not allow an immediate comparison of one pollutant with a different pollutant in terms of numbers of molecules but an atmosphere containing 1 ppmv sulphur dioxide contains the same number of pollutant molecules as one containing 1 ppmv ozone or nitrogen dioxide because the molecular weight and the volume of gas containing one gram molecule has already been taken into account. For this reason alone I have used the units ppmv or ppbv throughout, preferring ppmv rather than ppm to indicate that it is a concentration by volume in a volume of gas. This avoids confusion because ppm, i.e. a mass of a unit mass, is often used instead of standard (SI) metric units to express pollutant or contaminant levels in solids during analysis. In body tissues, for example, 1 ppm strychnine means 1 mg of strychnine per kilogram of body weight to a pathologist.

In some cases, ppmv or ppbv expressions of atmospheric concentration are entirely inappropriate, e.g. particulates or aerosols. Volume per unit volume measurements are, in such cases, impossible and mass per unit volume measurements must be qualified by the ranges of the particle sizes or the droplet diameters.

1.3 LEVELS AND LIMITS

We only have one world with one biosphere and one atmosphere. For centuries in England, Common Law has protected individuals from point sources of air pollution. In 1691, for example, the baker next door to Thomas Legg of London was made to put up a chimney 'soe high as to convey the smoake clear of the topps of the houses'. But when point sources become larger, more numerous and merge into polluted environments, air as well as individuals has to be protected by government using legislation. When pollution from one national state is alleged to cause problems in another country then international agreement is required.

Sources of air pollution are diverse and it is often almost impossible clearly to discriminate between them. Domestic sources often

Table 1.1 Representative apportionments of generated atmospheric pollution (based on 1968 figures for the USA)

Operation	Particulates (%)	Sulphur dioxide (%)	Oxides of nitrogen (%)	Carbon monoxide (%)	Hydrocarbons (%)
Power and heat	47.9	74.9	53.2	2.2	2.3
Transport	6.5	2.3	42.7	76.7	59.7
Refuse disposal	5.3	0.3	2.9	9.4	5.9
Other operations	40.2	22.4	1.1	11.7	16.6
Solvent evaporation	—	—	—	—	15.4

surround industrial sources and power or heat generation are required by both. Private individuals also contribute to the atmospheric problems that arise from various forms of transport or the disposal of rubbish. Table 1.1 attempts to show approximate proportions of pollution that are caused by particular activities without making the distinction between industrial and domestic apportionments. Owing to the burning of fossil fuels, power generation contributes significantly to the emission of particulates, sulphur dioxide and oxides of nitrogen but alternative fossil fuels contribute different pollutants. Table 1.2 shows typical ranges of stack pollutant gases emitted by power plants which have not been fitted with flue gas denitrification and desulphurisation.

Table 1.2 Typical ranges of gases (in ppmv) emerging from the stacks of power plants using different forms of fuel

Pollutants (ppmv)	Fossil fuel			
	Coal	Oil	Gas	Peat
SO_2[†]	400–3000	500–3000	—	100–1000
NO	300–1300	250–1300	200–500	100–1000
NO_2	10–40	10–40_	10–30	3–30
CO	—	—	100	2000–5000
Hydrocarbons	10–60	100–2000	100–2000	30
HCl	80	—	—	—

[†] SO_3 values are between 1–10 % of the amounts of SO_2.

Nearly every developed country has introduced legislation to control or limit emissions of atmospheric pollution and it is not within the scope of this book to consider individual examples of how it operates. However, international agreements do exist concerning desirable air quality standards. The European Economic Community for example uses both Limit Values and Guide Values. The difference between the two is that Limit Values have to be enforced by national legislation whilst Guide Values have no legal implication but indicate that European Economic Community governments should work towards achieving these levels. On top of these are air quality Guidelines recommended by World Health Organisation which are to be regarded as desirable objectives. Table 1.3 attempts to summarise these three types of international recommendations.

In the work place, standards based upon exposure limits or Threshold Limit Values (TLVs) are set in the USA for specific substances (including air pollutants) to which (it is believed) the majority of working adults may be exposed for five 8-hour working periods per week without showing ill effects. Countries of the European Economic Community and other developed countries usually abide by these Threshold Limit Values which were originally established by the American Conference of Government Industrial Hygienists (ACGIH) on the basis of animal experiments, medical

Table 1.3 Examples of international air quality standards for sulphur dioxide issued by the European Economic Community (EEC) and the World Health Organisation (WHO)

Directives and recommendations	SO$_2$	
	(μg m^{-3})	(ppbv)
EEC Limit Values (1980)		
Year (median of daily values)	120[†]	45
Winter (median of daily values)	180[‡]	68
(October–March)		
EEC Guide Values (1980)		
Year (daily arithmetic mean)	40–60	15–23
24 hour (daily mean)	100–150	38–56
WHO Guidelines (1985)		
Year (arithmetic mean)	30[§]	11

[†] 80 μg m^{-3} if smoke more than 40 μg m^{-3}.
[‡] 130 μg m^{-3} if smoke more than 60 μg m^{-3}.
[§] Equivalent figures for nitrogen dioxide and ozone are 30 and 60 μg m^{-3}, respectively.

Table 1.4   Threshold Limit Values (TLVs[†]; sometimes called Maximum Allowable Exposures, MAEs) for five 8-hour exposures (as a weighted average) per week for healthy humans[‡]

Pollutant	TLVs (ppmv)	TLVs (mg m^{-3})
Sulphur dioxide	5	15
Nitric oxide	25	30
Other oxides of nitrogen	5	—
Ammonia	50	35
Ozone	0.01	0.2
Carbon monoxide	50	55
Carbon dioxide	5000	9800
Carbon disulphide	20	68
Hydrogen fluoride	3	2
Fluorine	1	2
Hydrogen sulphide	10	15
Hydrocarbons	500	—
Peroxyacetyl nitrate	0.08	—

[†] Occupational Exposure Limits (OELs) in the UK.
[‡] Originally they were recommended by the American Conference of Government Industrial Hygenists (ACGIH) and are now the responsibility of the Occupational Safety and Health Administration (OSHA).

knowledge, epidemiology and other environmental studies. In the United Kingdom, for example, Threshold Limit Values are known as Occupational Exposure Limits or OELs. A number of these Threshold Limit Values (or OELs) are shown in Table 1.4. It may be quickly appreciated that the Threshold Limit Value for sulphur dioxide, for example, is well in excess of those recommended as international air quality standards (Table 1.3). This is because Threshold Limit Values were only designed to be applicable to workers over relatively short periods during their working life. On the other hand, air quality standards have to encompass long-term effects upon humans, vegetation, materials and the environment as a whole.

Irrespective of which are set, air quality standards do not of themselves create cleaner air unless policies are adopted to ensure that they are met. Implementation of controls and monitoring, inspections and enforcement must be rigorous if success is to be achieved. The strictest standards are useless if there is no intention to ensure their adoption and use in the environment, in the home or at the workplace.

1.4   DETECTION AND INDICATORS

Monitoring of air pollutants is a prerequisite of air quality control and may be done by a wide variety of methods which have widely differing sensitivities and specificities. Some are continuous whilst others require batch or 'grab' samples collected over a period, a few of which have to be concentrated before analysis. The continuous methods are more adaptable to automatic unattended operation but are usually more expensive and cannot be liberally employed over extensive areas for long periods. The period of automatic sampling by such continuous techniques has also to take into account the time taken for the detectors to respond to the pollutant. Cheaper non-continuous methods are usually standardised to fixed period of sampling but may also be semiautomatic although they often require manual operations during or after sampling. That is not to say the automatic continuous methods do not require attention; several of them require supplies of other gases which have to be checked and have moving parts, e.g. pumps, which need regular maintenance. Table 1.5 lists a number of continuous techniques for a range of pollutants within minimum levels of pollutant they may detect and the time taken for a 90 % response to a concentration change.

Chemiluminescence is a good example of a continuous technique which is used for both the detection of oxides of nitrogen and ozone. In the former, excited (*) nitrogen dioxide is formed from nitric oxide

Table 1.5   Continuous methods of air pollutant measurement

Pollutant	Technique	Response time	Detection limit
SO_2	H_2O_2/conductivity	3 min	10 ppbv
	Flame photometric	25 sec	0.5 ppbv
	Pulsed fluorescence	2 min	0.5 ppbv
NO	Chemiluminescence with O_3	1 sec	0.5 ppbv
NO_2	Reduction/chemiluminescence[†]	1 sec	0.5 ppbv
O_3	KI oxidation/electrolysis	1 min	10 ppbv
	Chemiluminescence with ethylene	3 sec	1 ppbv
	Ultraviolet spectroscopy	30 sec	3 ppbv
CO	Electrochemical	25 sec	1 ppmv
	Non-dispersive infrared	5 sec	0.5 ppmv
Hydrocarbons	Flame ionisation	0.5 sec	10 ppbv
	Non-dispersive infrared	5 sec	1 ppmv

† May also be used for NH_3 but reduction carried out at higher temperatures than 650 °C.

in the presence of excess ozone (Reaction 1.1) which emits light at 1200 nm (Reaction 1.2). Nitrogen dioxide may be converted to nitric oxide by heat (650 °C) prior to measurement and determined by difference from measurement of nitric oxide alone or in combination as total oxides of nitrogen:

$$NO + O_3 \longrightarrow NO_2^* + O_2 \tag{1.1}$$

$$NO_2^* \longrightarrow NO_2 + light \tag{1.2}$$

Ozone will also react with hydrocarbons such as ethylene (now called ethene) to form a light-emitting free-radical (see Appendix, Sect. 8.2) which decays to formaldehyde and emits light around 435 nm (Reaction 1.3) so it does not interfere with any light from the reaction of ozone with nitric oxide.

$$O_3 + C_2H_4 \longrightarrow 2CH_2O^{\bullet} \longrightarrow 2CH_2O + light \tag{1.3}$$

Older batch or semiautomatic techniques have been much improved with time. Table 1.6 lists a number of the more successful of these techniques, the period most often used for collecting the sample and the lowest detection limit in each case. Nevertheless they are more prone to interference by other contaminants. However, the once-popular West–Gaecke method for sulphur dioxide, which relied upon the formation of disulphitomercurate from tetrachloromercurate (Reaction 1.4) which then reacted with acidic *para*-rosanilinemethylsulphonic acid to give a pink colour, was refined to the point where possible interfering substances were almost eliminated:

$$[HgCl_4]^{2-} + 2SO_2 + 2H_2O \longrightarrow [Hg(SO_3)_2]^{2-} + 4Cl^- + 4H^+ \tag{1.4}$$

On the other hand, the equivalent Saltzmann procedure for nitrogen dioxide adopted for a time in the USA as a reference standard, whereby the nitrite formed from nitrogen in solution reacted to give a coloured diazo derivative of sulphite which could be stabilised with *N*-(1-naphthyl)ethylenediamine, had a number of drawbacks (including variable stoichiometry) which caused it to fall from favour. Similar problems have also beset the once-standard wet chemistry method for ozone and total oxidants. Originally, it was thought that when ozone reduced iodide to iodine this reaction always followed a stoichiometry of I_2 to O_2 of unity (Reaction 1.5). The subsequent

$$O_3 + 2H^+ + 2I^- \longrightarrow I_2 + H_2O + O_2 \tag{1.5}$$

$$KI + I_2 \longrightarrow KI_3 \tag{1.6}$$

formation of triiodide (Reaction 1.6) was then monitored by the spectrophotometric increase in absorbance at 352 nm, or the disappearance of iodide by ion-selective electrodes or the amperometric conversion of the triiodide. However, Reaction 1.5 is probably more complicated

Table 1.6 Non-continuous or semiautomatic methods of air pollutant measurement

Pollutant	Technique	Collection period	Detection limit
SO$_2$	H$_2$O$_2$/acid-base titration	24 h	2 ppbv
SO$_2$	H$_2$O$_2$/acid–base titration	24 h	2 ppbv
	West–Gaeke	15 min	10 ppbv
NO$_2$	Modified diazotisation	30 min	5 ppbv
O$_3$[†]	KI oxidation/spectrophotometry	30 min	10 ppbv
Peroxyacyl nitrates	[§]GLC/electron capture	—	1 ppbv
CO	Methanation/flame ionisation	—	10 ppbv
Hydrocarbons	GLC/flame ionisation	[‡]	1 ppbv

[†] Including total oxidants as well.
[‡] Samples require concentration prior to injection.
[§] GLC, gas–liquid chromatography.

than shown and a stoichiometric ratio of I$_2$ to O$_2$ of 1.5 has been observed, thus throwing doubt upon the accuracy by which ozone was being determined.

Chemical monitoring of air pollutants can sometimes be replaced by biological indicators. Certain lichens, for example, have long been known to be sensitive to atmospheric pollutants such as sulphur dioxide, hydrogen fluoride, ozone and peroxyacyl nitrates. Pollution mapping based on the distribution and abundance of lichen species is frequently undertaken but a number of precautions have to be taken to ensure reliability and relevance. For example, if lichens growing on trees (epiphytic lichens) are chosen, then the sampling has to be from the same species of tree growing in similar environmental conditions in the absence of unusual nutrients, herbicides or pesticides, etc. A great deal then rests on the experience of the survey team if valuable results are to be obtained and overlap between different pollutants and differences in climate (e.g. humidity) are to be avoided. Such surveys are used to best advantage when an unpolluted area is about to have an industrial plant located nearby. Surveys before and after may quickly show evidence of a deleterious change which requires corrective action.

Higher plants may also be used as bioindicators especially if they are very sensitive to one particular pollutant and show clear evidence of visible damage which is of a characteristic nature. A great deal of effort has been devoted to the development of a range of plants each of which show specific damage to one pollutant or accumulate the products of pollution in a characteristic manner. A list of those

Table 1.7 List of plant species suitable for use as air pollution bioindicators in Europe (after Steubing and Jager, 1982)[†]

Pollutant	Common name	Species and variety
Sulphur dioxide	Alfalfa, lucerne	*Medicago sativa* L. cv. Du Puits
	Clover	*Trifolium incarnatum* L.
	Pea	*Pisum sativum* L.
	Buckwheat	*Fagopyrum esculentum* Moench.
	Great plantain	*Plantago major* L.
Nitrogen dioxide	Wild celery	*Apium graveolens* L.
		Petunia sp.
	Ornamental tobacco	*Nicotiana glutinosa* L.
Ozone	Tobacco	*Nicotiana tabacum* L. cv. Bel W$_3$
Peroxyacyl nitrates	Small nettle	*Urtica urens* L.
	Annual meadow grass	*Poa annua* L.
Hydrogen fluoride and fluorides		*Gladiolus gandavensis* L. cv. Snow Princess
	Tulip	*Tulipa gesneriana* L. cv. Blue Parrot
General accumulators	Italian rye grass	*Lolium multiflorum* Lam. ssp. *italicum*
	Cabbage	*Brassica oleracea* L. var Acephala
Bark accumulators	Rose	*Rosa rugosa* Thunb.
		Thuga orientalis L.

† See selected bibliographies at the end of each chapter for source details.

indicator plants suitable for Europe is shown in Table 1.7. These plants may be container-grown under clean-air conditions to a predetermined condition and age and then taken to various locations over a whole area. They are subsquently collected after a fixed period of exposure, brought back to a central location and compared with equivalent clean-air controls. From the visible differences an assessment of air pollution conditions may be made. Whole countries like the Netherlands are covered by such networks which allow the integration of biological and chemical monitoring data.

This controlled exposure of bioindicators and comparison with equivalent clean-air controls is infinitely superior to subjective judgements of visible injury to pre-existing vegetation consisting of widely different species. The latter may only draw attention to the possibility of problems which may or may not be associated with air pollution damage. So many other factors like drought, insect attack, water-logging, fungal infection and temperature can cause visible injury to vegetation and, consequently, wrong conclusions may so easily be drawn. Even if bringing expensive chemical monitoring into a suspect locality is out of the question, the setting out of a few groups of sensitive plants often provides useful definitive answers without leading prematurely to heated debate.

Other bioindicators are possible and as yet are little used. Leaf-eating insects, for example, can distinguish leaves from polluted plants from those not exposed even though there are no visible differences between the leaves. Frequently, they prefer to attack the polluted leaves. Similarly, the microbial population on the surface of leaves changes in accordance with the atmospheric conditions and any alter-ations in the cuticular waxes on the leaf surfaces as a consequence of atmospheric pollution. In an interesting survey of atmospheric pollution being carried out by Irish schoolchildren, one of the tests follows the increase in numbers of pink yeasts after plating out leaf imprints. Simple though these tests using microbes and insects may be, there is still little understanding of the fundamental mechanisms underneath them. However, if tests like these could be devised and made, semiquantitative then useful survey information could be gained without the extensive use of expensive monitoring equipment.

1.5 THRESHOLDS AND INJURY

One of the classical assessments of toxicology is to determine the median concentration which is lethal to a certain proportion of organ-isms. The LC_{50} or LD_{50} are most frequently used in the testing of drugs, etc. These are the lethal concentrations or doses that kill 50 % of the test animals in a certain period (i.e. 48 hours). These are then applied to the mid-point of sigmoidal curves (Fig. 1.2) and then back-projected to give notional 'safe' concentrations or doses that are said to be harmless or only kill an 'acceptable' fraction of the population. Apart from providing a quick screening of likely hazards from an unknown toxicant, the logic of the rest of the interpretation is questionable. Death and sublethal harm are often unrelated. Even assuming all variation (which is usually considerable) has been removed from such experimentation, there is no direct means of extrapolating what happens, for example, in a rat to a human. Other variations on the same theme such as the MLD (minimum lethal dose) or MDNF (maximum dose never fatal) are little better.

As lethality is only a very crude measure of toxicity, possible sublethal effects have been monitored by toxicologists. The problem still remains: 'What change is harmful and what is within the normal

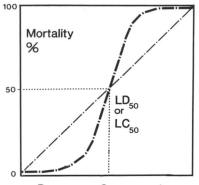

Dosage or Concentration ⟶

Fig. 1.2 The concept of the LD_{50} or LC_{50} (lethal dose/concentration that kills 50 % of a group) normally assumes a sigmoidal mortality curve which allows the LD_{50}/LC_{50} to be back-projected to a dose or concentration where only a statistically small number of individuals are affected. In the case of some substances, the sigmoidal nature of response is scarcely evident (lighter plot) so that no 'safe' dose or concentration exists.

range of homeostasis (i.e. the internal self-regulation of cells or organisms)?'. Numerous parameters of sublethal toxicity (e.g. biochemical, mutagenic, carcinogenic, teratogenic) are used. For each toxic compound there is usually a different type of response curve, some linear and others with thresholds. Each study is different and the variables are considerable especially when air pollutants are considered. The Threshold Limit Values set in the USA were derived by a combination of lethal and sublethal studies. They have practical value at the work place for healthy individuals but critical groups – the young, the old, the ill and the pregnant – are not covered because the original objectives whilst setting these values reflected an active and largely male working population and never included the whole population.

Ideas and concepts developed by animal toxicologists have been picked up by plant pathologists who perform an equivalent function with respect to vegetation. Pathological change in plants usually involves structural damage and this is frequently shown as visible injury to leaves. By appreciating the fact that damage may not always be visible the concept of 'invisible injury' has been developed. This type of unseen injury is thought to occur outside the normal ranges of homeostasis and to give rise to reductions of growth which, if prolonged, would ultimately lead to visible injury. The idea of at least two thresholds for air pollutants, one at which no effects merge into some change, and another at which reversible effects became irre-

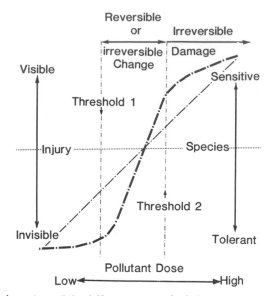

Fig. 1.3 Adaptation of the LD_{50} concept which has been used for the interpretation of plant effects. The sigmoidal response is said to give rise to 'thresholds' so that at certain concentrations or doses a response is given which may be either reversible or irreversible. Because some of these have not given rise to visible symptoms yet they inhibit growth, the idea of 'invisible' injury being caused has been popular.

versible, is often advocated. For these to exist, sigmoidal curves like that shown as a heavier line in Fig. 1.3. have to be demonstrated. This has rarely been demonstrated during studies of effects of air pollutants upon plants in practice. Indeed, it is very difficult to show a clear linear relationship between dose or concentration and some degree of change. Nor has any real distinction been made between so-called 'invisible injury' and the wide ranges of normal homeostasis. As Fig. 1.3 also attempts to show, reversible and irreversible changes may occur simultaneously. Response will always be different even between individual plants in the same population. Some will tolerate change and affect homeostasis (see Preface) immediately afterwards whilst others may be sensitive and be unable to adjust to the altered conditions. The concept of damage is best viewed from an energetic consideration. Everything has a price just as in daily life. The bio-energetic cost to a plant of normal homeostasis in order to cope with external influences such as atmospheric pollutants may cause losses of energy which then cannot be converted into growth, reproduction, etc. The further along the response curve of Fig. 1.3 (be it a heavy or a lightly dashed and dotted line) means the more energy that has to be devoted to maintenance. At some point a negative balance between income and expenditure (the plant equivalent of bankruptcy) is achieved and this leads later to visible injury. In other words, there is no sharp distinction between normal homeostasis and so-called 'invisible injury'. One counterbalances the other in a cellular profit or loss (or cost–benefit assessment) that is constantly being updated as the surrounding environment changes. This may appear to some as simplistic or even heretical but, as the rest of this book goes on to show, the primary attack of atmospheric pollutants is often upon the powerhouses of the cell: the mitochondria, producing energy in the form of adenosine triphosphate (ATP) by oxidising carbon sugars, and the chloroplasts which harness light to form ATP and reducing power which is then used to fix carbon dioxide.

Those who are not familiar with the basic mechanisms (i.e. photosynthesis and respiratory phosphorylation) involved in these two organelles may find benefit in the short basic descriptions included in the Appendix (Sect. 8.3 and 8.4) before reading any of the later chapters.

1.6 COMPOSITION AND CLIMATE

'A solid has volume and shape; a liquid has volume but no shape; and a gas has neither volume nor shape', wrote Lodge, meaning that gas or vapour is matter in a perfectly fluid state. The whole atmosphere behaves, more or less, in a similar manner and consists of a mixture of gases (i.e. nitrogen 78.08 %, oxygen 20.95 %, argon 0.93 %, carbon dioxide 0.035 %, neon 0.0018 %, helium 0.005 % and krypton 0.0001 %) extending to about 1000 km above the Earth's surface at the Equator and rather less (800 km) at the Poles. The total mass of the atmosphere is 5.14 Pt (where P, peta, is 10^{15} and a metric tonne, t is 1000 kg) — far smaller than the solid or aqueous components of the planet and more than 99.9 % of that mass of gas is below an altitude of 50 m.

There are considerable temperature gradients in the atmosphere with distinct warm and cold layers above the surface of the Earth (Fig. 1.4). These are referred to as the 'troposphere' (closest to the surface) where there is a decrease of temperatures with height up to 12 km, the 'stratosphere, (12–50 km) where temperatures increase with height, the 'mesosphere' (50–80 km) where they decrease again, and then in the 'thermosphere' (80 km upwards) they rise once more, sharply so after 110 km. Inversions of temperature between these different regions are called the 'tropopause', the 'stratopause' and the 'mesopause', respectively (Fig. 1.4). Temperature changes are brought

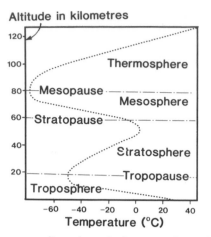

Fig. 1.4 Temperature gradient through the atmosphere showing the nomenclature of the different regions and the heights at which they occur.

about by the absorption of visible, ultraviolet and near infrared solar radiation and the re-emission and absorption of thermal radiation from the Earth by various atmospheric gases, especially carbon dioxide, ozone and water vapour. The concentrations of these constituents of the atmosphere, as might be deduced from Fig. 1.4, are not uniform with altitude. Figure 1.5 shows their typical concentrations (or mixing ratios) in the troposphere and lower stratosphere. From a consideration of atmospheric pollution, the part played by carbon dioxide is important. Whilst it does not absorb incoming solar radiation to any extent, it strongly absorbs a band of infrared from the Earth's thermal radiation preventing loss to outer space. This heating blanket has been called the 'greenhouse effect'; the consequences of which are considered elsewhere (Sect. 6.2a).

Air is also warmed by combustion processes on Earth as well as by absorbtion of reflected or re-emitted thermal radiation. In some parts of the populated world this may be as high as 20 % of the total incoming solar radiation. As the temperature of the air near the surface rises, its density falls causing it to expand and rise. Heat is lost to the higher but cooler regions of the troposphere cause it to fall once again. This gives an up-and-down movement to masses of tropospheric air. Vertical velocities may reach 20 m sec^{-1} in thunderstorms but 10 cm sec^{-1} are more normal. Pollutants are moved sideways by wind systems that mainly blow from west-to-east in the middle latitudes of both hemispheres. At 30°N, for example, a 35 m sec^{-1} west-to-east flow takes about 12 days to get right round the world. North–south oscillations also occur at the same time as

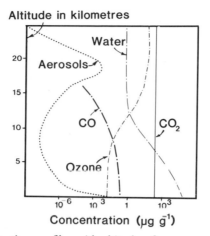

Fig. 1.5 Concentration profiles with altitude of some components of the atmosphere showing the rise of ozone and aerosols towards the tropopause and the steady fall of water and carbon monoxide with increasing height. Carbon dioxide, by contrast remains at steady concentration when expressed on a mass per unit mass, basis.

west–east flows which, when combined with up-and-down movements, give a turbulency at low altitudes called 'Hadley-cell circulation'. This is a dominant feature at the Equator but at middle latitudes (where most atmospheric pollution is emitted), north–south eddies in west–east prevailing winds are the most important features. Exchanges between the troposphere and the stratosphere take place mainly at the Equator as warm air masses rise over the tropics but are prevented from rising beyond that because temperatures rise with altitude above the tropopause. This means that traces of pollutants can remain in the equatorial stratosphere for much longer (often years) and often move either north or south before mixing and transfer back to the troposphere occurs. This mechanism is virtually the only one by which atmospheric pollutants are transferred between hemispheres.

The troposphere also contains varying amounts of water vapour (Fig. 1.5); the warmer it is, the more water vapour that may be retained. If the troposhere is saturated and then cooled, the water condenses to form clouds, mists or fogs. Mists differ from fogs by virtue of having smaller droplet sizes (less than 2 μm). Water vapour produced by combustion is an insignificant component of the global tropospheric water or hydrological cycle (Fig. 1.6) but fogs are often associated with urban and industrial areas because the associated particulates pollution provides the necessary nuclei for the condensation of water vapour especially at night when temperatures are lower. The lower stratosphere, by contrast, can hold only minute amounts of moisture because as water vapour enters the tropical tropopause it is frozen out at temperatures approaching −80°C to form high cirrus cloud.

The annual global precipitation rate over the land is about 71 cm y^{-1} whilst evaporation rates are much lower (47 cm y^{-1}). Over the oceans, which have more than double the land area, these values are reversed. Global precipitation over the sea amounts to 110 cm y^{-1} whilst evaporation is equivalent to 120 cm y^{-1}. This means there is a net transfer of 44.5 Tt of water (where T, tera, is 10^{12}; t, a metric tonne, is 1000 kg) from the oceans to the land surface which is compensated by an equivalent amount of run-off from the land to the sea. However, because the atmosphere is such a small reservoir by comparison to the land or the oceans, particularly for water, transfer through the atmosphere must be very quick.

The residence time of a substance (S) in a reservoir is equivalent to the total capacity of S divided by the input or the outflow of S. In the case of water in the atmosphere (taking the values shown in Fig. 1.6), total capacity (1.3 Tt) divided by net flow from oceans-to-land (45 Tt y^{-1}) gives 0.03 years or just under 11 days as the residence time for water vapour in the atmosphere. By comparison to other residence times calculated for other atmospheric gases (Table 1.8) this is a very short time indeed and similar to that of sulphur dioxide. Only those gases like carbon dioxide which have residence times much longer than half a year have sufficient mixing to give uniform global concentrations.

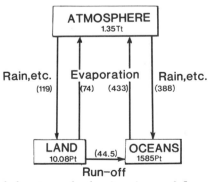

Fig. 1.6 The global water cycle showing sizes and fluxes of water between the ocean, the atmosphere and the land surfaces. The pool size units are given in Tt (tera, 10^{12}) or Pt (peta, 10^{15}; t, metric tonnes, is 1000 kg) and the fluxes are in Tt y^{-1}. There is a net transfer of water from the oceans to the land via the atmosphere of 44.5 Tt y^{-1} which balances the quantities lost by run-off.

Table 1.8 Approximate residence times for atmospheric gases in the atmosphere

Component	Time (years)
Ammonia, hydrogen sulphide	less than 0.005[†]
Nitric oxide, nitrogen dioxide	less than 0.01[†]
Sulphur dioxide	less than 0.02[†]
Water	0.03
Ozone	0.3
Carbon monoxide	0.4
Methane	3
Nitrous oxide	7
Carbon dioxide	10
Oxygen	8000

[†] Highly variable depending upon moisture content and a variety of other factors.

1.7 PARTICLES AND SUSPENDED MATTER

Particulates in the air consist of very small solid- or liquid-suspended droplets which have a number of different names. Each have quite specific meanings despite being popularly used in an interchangeable manner. Table 1.9 attempts to characterise a number of these by nature and size. In the case of grits and most coarse dusts, particles with diameters greater than 50 μm are visible and those greater than

Table 1.9 Different types of suspended particulates in the atmosphere

Name[†]	Nature	Size (diameter) (μm)
Grit	Solids, settle out quickly	Greater than 500
Dust	Solids, settling more slowly	2–500
Smoke	Gas-borne solids	Less than 2
Mist	Liquid droplets	0.1–2
Aitken nuclei	Solid or liquid droplets	0.003–0.1

[†] The term 'aerosol' appears to have a number of different interpretations. Aerosols may consist of solid or liquid droplets which are either less than 1 μm in diameter, but also include those mechanically generated particles with a diameter greater than 1 μm which remain suspended for long periods. If the particles collectively obscure visibility they are said to form a 'haze'.

10 μm quickly settle close to the point of pollution. Those smaller with diameters than this may remain suspended for some time and if very small (0.1–2 μm) can act as nuclei for the condensation of water to form clouds and are either removed as it rains (by a process known as 'rain-out') or hit by falling rain ('wash-out'). They can remain suspended for very much longer if they escape the lower troposphere to upper altitudes where they may remain for months or even years. Extremely small particles (0.005–0.1 μm) are known as 'Aitken nuclei' and are formed primarily as condensation products from hot vapours which aggregate after a series of chain reactions. These Aitken nuclei may then coagulate together to form larger particles (0.1–2 μm) which are gradually rained-out or washed-out. Vapours of low volatility (e.g. hydrocarbons) may assist with the condensation and growth of these larger nuclei.

As might be expected, particulates contain a wide diversity of water-soluble and insoluble components. The latter consists of elemental and organic carbon, iron oxides and a variety of other materials such as rock-derived powders (e.g. quartz, mica, china clay and fibres such as asbestos), all of which are known to cause industrial dust diseases like pneumoconiosis, silicosis and asbestosis. Toxic metals (e.g. lead or cadmium) may also be minor components of particulates which have their own toxicological problems.

Soluble components of particulates include the common cations (sodium, potassium, calcium, magnesium, ammonium, protons) as well as chloride, sulphate and nitrate. Sodium, chloride and magnesium mainly arise naturally from sea-spray whilst calcium and potassium are similarly derived from soil-blown material. Ammonium, sulphate, nitrate and protons are produced by natural (e.g. volcanoes) or man-induced emissions into the atmosphere of ammonia or the oxidation of sulphur dioxide, nitric oxide, etc. Reductions in visibility due to aerosol hazes is often claimed to be caused by combinations of ammonia with oxidation products of atmospheric pollutants to form reflective particles of NH_4HSO_4, $(NH_4)_2SO_4$ and NH_4NO_3. Particulates also absorb and reflect incoming solar radiation. In periods before smoke-control legislation as much as 50 % of the incoming light to big cities like London could be lost in the winter (10 % in summer). Even now there may be as much as an average 15 % yearly loss plus an increased likelihood (up to 10 %) of cloud cover because of the nucleating effect of small particulates during the formation of clouds. Larger particulates falling quickly to Earth may accelerate the atmospheric pollution-induced corrosion of materials such as stone-work and by adhering to the surface of leaves may cause a further reduction in net photosynthesis by blocking the stomata or by reflecting or absorbing incident light energy.

Because of their diversity of sizes and composition, particulates are

Table 1.10   Annual atmospheric deposit in UK towns in the early part of the twentieth century in metric tonnes per square kilometre (after Ashworth, 1933)

Year (April–March)	London	Glasgow	Newcastle	Rochdale	Malvern
1916–1917[†]	157	149	279	292	35
1917–1918[†]	149	175	249	417	31
1920–1921	120	126	199	298	30
1921–1922	111	100	211	203	27

[†] The war-time increase in extra pollution is equivalent to what might have been expected if one and a half extra days were being worked, which with overtime was indeed the case.

difficult to monitor and define. Most common estimates rely upon gravimetric measurements of accumulated material collected by filtration to give 'total suspended particulates'. Standards vary from country to country but the air quality standard set for particles in the USA is 75 μg m^{-3} as an annual mean and 260 μg m^{-3} as a daily or 24-hour average not to be exceeded for more than one 24-hour period per year. Another way of measuring smoke particulates is to determine the staining capacity of air upon filters. Limit values (enforceable) of smoke set by the EEC are 80 μg m^{-3} (as a median of daily mean), 130 μg m^{-3} (as a median over the winter months October to March) and 250 μg m^{-3} (as the 98th percentile of daily values). Guide values (to which EEC members should aim) are 40–60 μg m^{-3} smoke as an arithmetic mean of daily values and from 100 to 150 μg m^{-3} as a daily mean. These and preceding regulations and recommendations have gone a long way towards cleaning the atmospheres above developed countries in terms of visibility and amenity. In retrospect it is often difficult for the young and the population at large in developed countries to appreciate just how bad particulate pollution was in the past or still is in certain parts of developing countries. In order to demonstrate this without using highly selected photographs of 'satanic' mills, figures have been reproduced in Table 1.10, gathered over the period 1916–1922 by Ashworth for London, Glasgow, Newcastle, Rochdale and Malvern in the UK. Rochdale was once one of the most important cotton-spinning towns in the world (though no longer so) and Malvern is a health resort to the south-west (and cleaner) side of Birmingham. From figures available now (which are not easily converted into equivalent quantities) the present depositions for Rochdale are less than those for Malvern then – which gives some idea of just how much improvement has been made.

SELECTED BIBLIOGRAPHY

Ashworth, J. R., *Smoke and the Atmosphere*. Manchester University Press, 1933, Manchester.

Dix, H. M., *Environmental Pollution*. John Wiley, 1981, Chichester.

Duffus, J. H., *Environmental Toxicology*. Edward Arnold, 1980, London.

Giffin, F., *Industrial Gases*. Frederick Muller, 1964, London.

Harrison, R. M., *Pollution: Causes, Effects and Control*. Special Publication no. 44, Royal Society of Chemistry, 1983, London.

Mellanby, K., *The Biology of Pollution*. Studies in Biology No. 38. Edward Arnold, 1972, London.

Raiswell, R. W., Brimblecombe, P., Dent, D. L. and Liss, P. S. *Environmental Chemistry*. Edward Arnold, 1980, London.

Readings from Scientific American, *Chemistry in the Environment*. W. H. Freeman, 1973, San Francisco, Calif.

Steubing, L. and Jager, H. J., *Monitoring of Air Pollutants by Plants: Methods and Problems*. Dr. W. Junk Pub., 1982, The Hague.

Chapter 2

Sulphur dioxide

'That knuckle-end of England –
that land of Calvin, oat-cakes and sulphur'[1]

by the Reverend Sydney Smith (1771–1845)
whilst commenting upon Scotland.

2.1 SOURCES AND CYCLING OF SULPHUR

(a) Man-made atmospheric changes

Both natural and man-made emissions contribute to atmospheric levels of sulphur dioxide (SO_2). A gas which has an unpleasant, highly irritating odour when it occurs in large quantities. Global emissions from natural sources due to microbial activity, volcanoes, sulphur springs, volatilisation, sea-spray and weathering processes (see Sect. 2.2b) amount to 128 Mt (M, mega, is 10^6; t, 1000 kg) of sulphur per annum. In 1968, these were only 50 % greater than those caused by the activities of man (65.2 Mt sulphur y^{-1}) although they were more widely distributed over the surface of the planet. By contrast, more than 90 % of the man-made emissions arise just from the urban and industrialised areas of Europe, North America, India and the Far

---

[1] The second insult alludes to the pollution problems of Edinburgh around this period which gave rise to the nickname 'Auld Reekie' for that city.

East. Such emissions of sulphur dioxide were highest (76.8 Mt sulphur y^{-1}) during the mid-1970s but have fallen over the last decade as a result of emission controls, changes in the patterns of fuel consumption and economic recession. Unfortunately, forecasts of increased coal consumption towards the year 2000 predict that global man-made emissions of sulphur dioxide could rise again by about 30 % upon present-day rates to 85 Mt of sulphur per annum.

The burning of coal contributes by far the greatest proportion of man-made emissions of sulphur dioxide (see Table 2.1). Extraction of elemental sulphur and iron pyrites (FeS) used for the production of fertilisers and other sulphur-containing compounds used by man is often called 'voluntary' sulphur production. These sources may be reduced in future in favour of the re-using of 'involuntary' by-products (e.g. gypsum, $CaSO_4$) from emission-reduction devices fitted to power stations. Such changes may take place because of economic factors rather than by legislative intervention.

Technical advances have also been made for the removal of sulphur from fuels before and during combustion, not just by extracting them from petroleum products but also by removal from pulverised coal although this latter process is very expensive. The recent development of fluidised-bed combustion involves the burning of small particles of solid or liquid fuel at relatively low temperatures (500–700 °C). These are held in a state of suspension by jets of air injected under pressure directed upwards from below through many inlets. Steam and

Table 2.1 Estimates of global exchange of sulphur in 1968 caused by the activity of man (mainly from Anderson, 1978)

Source	Consumption $(Mt\,y^{-1})$	Recovery $(Mt\,y^{-1})$
Voluntary		
Elemental sulphur extraction	18.0	18.0
Iron pyrites extraction	11.0	11.0
Involuntary		
Copper smelting	5.8	
Lead smelting	0.7	
Zinc smelting	0.6	
Oil refining	16.9	22.0
Oil burning	10.3	
Coal burning	46.0	
Natural gas burning	6.7	
Total	116.2	51.0

water pipes immersed in the 'bed' transfer the heat to turbines very efficiently. The potential amount of sulphur dioxide emitted by combustion normally varies according to the sulphur content of the fuel. However, limestone or dolomite particles can be added to the fluidised 'bed' to cause a reduction of sulphur dioxide by up to 90 % because the sulphur is trapped in the form of sulphates of calcium or magnesium within the ash. This recent technique thus adds to the amount of 'involuntary' sulphur that may be recovered and allows coal of high sulphur content to be used in the presence of stringent emission controls. The low temperature of combustion also reduces the emissions of oxides of nitrogen (Ch. 3) and carbon monoxide (Ch. 6) because of the air-saturating conditions. Of all the combustion technologies used for power generation, fluidised-bed combustion would appear to have many advantages in its favour for the future. Table 2.2 lists a number of features of the different methods of power generation by combustion which illustrate this potential.

Table 2.2 Efficiencies and emission problems of power generation by combustion processes

Type of power plant	Maximum thermal efficiency (%)	Emissions		
		CO	SO_2	NO_x
Oil-fired	35–37	trace	high	considerable
Natural gas- fired	35–37	trace	low	very high
Conventional coal-fired	35–40	low	high	considerable
Gas Turbine	18–28	high	very high	very high
Fluidised bed	50–70	trace	low	low

Most man-made emissions of sulphur dioxide are made through tall stacks injecting the gases into the troposphere to heights at which the temperatures of the surrounding air and the rapidly cooled stack gases match. A discrete plume is then often carried away at that level. Naturally, the behaviour of the plume is dependent upon the prevailing weather features and the contents may be either dry deposited within 1 or 2 days close by or carried many hundreds of kilometres away from the source. When meeting wet conditions at any time on this journey, sulphur dioxide (and oxides of nitrogen) may be removed by wet deposition (acid rain). Such wet-deposition processes, and their consequences, are separately considered in their own right in Ch. 4 but at least four processes of dry-removal of sulphur dioxide have been predicted to take place in the atmosphere without the immediate presence of water droplets. All four processes are subsequently capable of generating acidic deposition because each is ultimately capable of forming sulphur trioxide (SO_3) which later reacts with water to form sulphuric acid (Reaction 2.1).

$$SO_3 + H_2O \rightleftharpoons H_2SO_4 \qquad (2.1)$$

Firstly, monatomic oxygen may react with sulphur dioxide (Reaction 2.2) within the plume, or even higher in the stratosphere. (Note: in Reaction 2.2 the letter M represents a third entity such as another gas molecule capable of carrying away excess energy associated with the formation of a new bond.) Less likely is a second mechanism based upon Reactions 2.3 and 2.4, which relies upon the presence of free-radical hydrocarbons which only occur to any great extent in very heavily polluted urban atmospheres. Thirdly, an alternative photo-chemical reaction to produce a highly energised form of sulphur dioxide (SO_2^*) which combines with oxygen to form sulphur tetroxide (SO_4) has been suggested. As sulphur tetroxide has never been positively identified in the atmosphere or under simulated conditions in the laboratory this light-driven acidification process is thought to be highly unlikely.

$$SO_2 + O + M \longrightarrow SO_3 + M \qquad (2.2)$$

$$O_3 + C_2H_2 \longrightarrow {}^{\bullet}CH_2OO + CH_2O \qquad (2.3)$$

$${}^{\bullet}CH_2OO + SO_2 \longrightarrow SO_3 + CH_2O \qquad (2.4)$$

The fourth possibility of dry, gas-phase removal of sulphur dioxide, and by far the most important, has been shown to involve highly reactive hydroxyl ($OH^{\bullet}$) free radicals (see Appendix, Sect. 18.2, for a fuller description) which are formed from either ozone (see Ch. 5) or monatomic oxygen (Reactions 2.5 and 2.6) molecules. These radicals readily react with sulphur dioxide (Reaction 2.7) to form $HSO_3^{\bullet}$ radicals which are then oxidised to peroxyl radicals ($HO_2^{\bullet}$) and sulphur trioxide (Reaction 2.8). An equivalent sequence of reactions may also form nitric acid (see Ch. 3) from nitrogen dioxide.

$$O_3 + light \longrightarrow O + O_2 \qquad (2.5)$$

$$O + H_2O \longrightarrow 2HO^{\bullet} \qquad (2.6)$$

$$OH^{\bullet} + SO_2 + M \longrightarrow HSO_3^{\bullet} + M \qquad (2.7)$$

$$HSO_3^{\bullet} + O_2 \longrightarrow HO_2^{\bullet} + SO_3 \qquad (2.8)$$

The important point emerging from the predominance of this mechanism of acidification involving hydroxyl radicals is that warm, bright conditions of summer, which favour ozone and monatomic oxygen

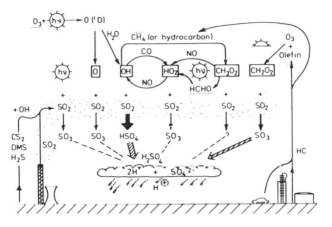

Fig. 2.1 Gas-phase oxidation and removal mechanisms for atmospheric sulphur dioxide (Courtesy of Drs. Cox and Penkett, AERE, Harwell, UK and D. Reidel. Publ. Co., Dordrecht, Netherlands.)

formation (and hence hydroxyl radicals) also promote acidification in the dry phase; whereas wet phase reactions (see Ch. 4) become more important in winter.

Figure 2.1 shows a number of the interrelationships between dry-removal mechanisms of sulphur dioxide discussed here and the corresponding gas–liquid-phase reactions which are treated separately in Ch. 4.

(b) Natural changes

Sulphur dioxide is emitted into the atmosphere by volcanoes and hot springs along with other gaseous forms of sulphur such as hydrogen sulphide (see Ch. 6). In terms of overall cycling of sulphur, however, weathering of sulphur-containing minerals such as gypsum, $CaSO_4$, is quantitatively more important. Such weathering is accelerated by acidification processes, be they caused by microbial activity or by atmospheric pollution. Later, most of this abraded sulphur slowly enters soils directly to act as a plant nutrient. In agricultural areas where soils are deficient in sulphur, it would normally be added artificially along with nitrogen, phosphate and potassium in fertilisers. However, in most regions of Europe and the eastern USA subjected to sulphur dioxide pollution, such treatments are quite unnecessary because the amount of sulphur falling on the land by wet and dry atmospheric deposition is more than four times that achieved by weathering even in areas well endowed with rocks with a high sulphur content.

Sea-spray is an alternative route by which sulphur compounds may enter the atmosphere. Such spray particles may remain suspended for long periods because the droplet sizes in this case are often very small. Most of this sulphur in the form of sulphate stays above the oceans and returns to the sea although about 10 % may be carried inland.

The global significance and interrelationships that exist between man-made, natural and microbial interconversions of sulphur between the buried sulphur reserves of the world, the surface of the land, the atmosphere and the oceans are illustrated in Fig. 2.2. The important features to note are that the amount of sulphur dioxide generated by the activities of man exceeds weathering by a factor of nearly 5. Furthermore, more of this atmospheric pollutant returns to the oceans rather than to the land because of the greater surface area of the sea. What is not quite so evident, but equally important, is that the temporary reduction (in geological terms) of sulphur reserves by the activities of man, leading to increases of sulphur in the atmosphere and made available to the biosphere, will eventually be reversed to end up as deposits on the floor of the oceans to form new sulphur reserves.

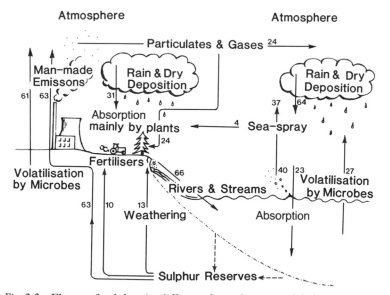

Fig. 2.2 Fluxes of sulphur in different forms between global compartments. The amounts (taken from Anderson, 1978) are expressed in Mt y^{-1} (megatonnes per annum; where 1000 kg is 1 metric tonne).

(c) Microbial changes

Land plants, seaweeds and microbes 'volatilise' sulphur into the atmosphere as a number of different compounds. Vegetation, for example, releases hydrogen sulphide as a mechanism for the elimination of excess sulphur (see Ch. 6) but microbes may use and release significant quantities of sulphur in a variety of different ways.

Some of the microbial exchange between the biosphere and the physical compartments illustrated in Fig. 2.2 are worth considering in their own right because they illustrate a diversity of different types of sulphur metabolism. This stems directly from the wide variety of individual chemical and energetic forms that sulphur may take up. By gaining an understanding of these differences, a more balanced appreciation of the interactions of sulphur dioxide or its derivatives with living systems may be subsequently gained.

Sulphur, like nitrogen, is unusual in that it may take a large number of oxidised or reduced states. Some of the important valency states of sulphur in different compounds are illustrated horizontally in Fig. 2.3. These compounds also differ energetically from each other, a concept best expressed in terms of their electrochemical potentials (shown vertically in Figure 2.3). Such a scale is standardised to the electrode potentials created by hydrogen and oxygen. These are

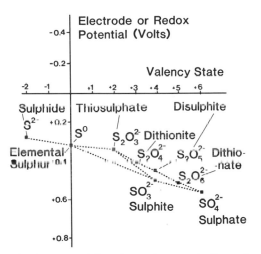

Fig. 2.3 Plot of redox potential against valency state for various sulphur compounds. The vertical axis has been turned upside down because electronegative compounds contain more energy and naturally donate electrons to compounds which are more electropositive. The scale of electrode or redox potentials represented in this way gives a better appreciation of the relative energies within different compounds.

dependent upon temperature and the degree of ionisation so when placed in biological context (25 °C and pH 7) they have the values of −0.42 V and + 0.8 V, respectively.

A firm grasp of the definitions associated with oxidation and reduction is invaluable. The process of reduction means 'the addition of electrons to an oxidised compound' (which may or may not include the addition of protons (H^+) at the same time). Conversely oxidation means 'the removal of electrons from a reduced compound'. The two concepts may be encapsulated in an idealised form as shown below (Reaction 2.9). Furthermore, as electrons must come from somewhere or be gained by another compound, it means 'for every reduction there must be a corresponding oxidation' and *vice versa*. Simple as this appears, it is the essence of an understanding of a number of apparently complex and various interactions that take place in living systems or outside them. Life forms have the ability to encourage the flow of electrons from a reduced form at a certain electropotential to an oxidised form at a lower potential and, as they flow, the chemical energy that is released may be trapped in other chemical forms or released as heat. This in turn allows other reactions to proceed faster.

$$\text{reduced form} \quad \underset{\text{reduction}}{\overset{\text{oxidation}}{\rightleftharpoons}} \quad \text{oxidised form} + \text{electron(s)} \qquad (2.9)$$

Figure 2.3 shows on the vertical axis the relative electrode potentials of the different commoner forms of sulphur at molar concentrations: the higher up the scale, the higher the potential. Unfortunately, the units and their signs originally adopted to denote electrical potential are confusing. Compounds at high negative potential or with high electronegativity are best placed upwards on such diagrams to denote their greater ability to donate electrons to lower potentials (or to more electropositive compounds). This scale of 'redox' potentials may be interpreted much better as a scale of relative energies. Thus as electrons flow downwards from electron donors (which become oxidised themselves) in Fig. 2.3 to electron acceptors (thereby reducing them) they release energy as they do so. Conversely, in order to make electrons flow in the opposite direction (upwards) energy has to be put in. The analogy with a series of waterfalls in an ornamental garden driven by an electric pump is most appropriate.

If, for example, electrons were allowed to flow from sulphite to sulphate, 715 kJ (joule) per mole (molecular weight in grams) of heat would be released but if they flow from sulphide to elemental sulphur only 213 kJ mol^{-1} would be released because elemental sulphur is more electronegative than sulphate. Such redox pairs are said to form redox couples (e.g. Reactions 2.10 or 2.11). A more electronegative couple such as S^0/S^{2-} will supply electrons to a more electropositive couple (e.g. SO_3^{2-}/SO_4^{2-}). This may be demonstrated practically if the two couples in separate solutions are connected or coupled electrically to each other. The potential of the more electronegative couple will be maintained as long as there is sulphide to be converted to sulphur and current will flow between the two systems.

$$S^{2-} \quad \rightleftharpoons \quad S^0 + 2e \qquad (2.10)$$

$$SO_3^{2-} \quad \rightleftharpoons \quad SO_4^{2-} + e \qquad (2.11)$$

Living systems also have distinctive organic couples which fit between such inorganic couples to permit the flow of electrons from more reduced couples (e.g. organic acids such as succinate/fumarate) or allow electrons to flow away from them to more electropositive couples (e.g. $\frac{1}{2}O_2/H_2O$). This ultimately allows other biochemical mechanisms to operate so that the energy within this flow of electrons is chemically trapped in a series of short steps rather than being lost as heat in one big rush.

If organic couples are placed between S^{2-}/SO_4^{2-} and $\frac{1}{2}O_2/H_2O$ as in members of the bacterial genera *Thiobacillus*, *Thiovulum* and *Thiospirillopsis*, then a process known as 'chemoautotrophy' (see Fig. 2.4) occurs. Trapped energy from this flow of electrons is then used to form adenosine triphosphate (ATP), the dominant energetic currency of living systems which can be used to assist with the

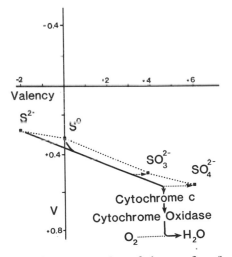

Fig. 2.4 Diagrammatic representation of electron flow (heavy line) during chemoautotrophy in the aerobic bacterial genera such as *Thiobacillus*, *Thiovulum* and *Thiospirillopsis*.

biosynthesis of virtually every other compound living organisms require. Chemoautotrophic bacteria tend to occur on the surface of stagnant ponds at the boundary of anaerobic (oxygen-lacking) conditions which are the source of the sulphide and aerobic (oxygen-rich) conditions which provide a final electron acceptor couple ($\frac{1}{2}O_2/H_2O$).

Some chemoautotrophs are responsible for an immense amount of economic damage to buildings because they can tolerate high acidity (below pH 1) caused by the sulphuric acid they release. The sulphuric acid released by *Thiobacillus concretivorus*, for example, steadily attacks the calcium carbonate of concrete, mortar or limestone thus rotting away the structure. Damage to sewers is particularly severe because the anaerobic conditions which prevail for long periods cause the build-up of reduced forms of sulphur which are then followed by surges of oxygenated water from downpours – ideal conditions for the growth of *Thiobacillus* and its concrete-rotting exudations.

More useful chemoautotrophic soil bacteria also appear to oxidise sulphur dioxide absorbed into the soil from the atmosphere into sulphate which then can be taken up by plant roots. Similarly, hydrogen sulphide released by anaerobic soils is detoxified by chemoautotrophic bacteria nearer to the surface of the soil before it can escape into the atmosphere to cause more of a problem (see Ch. 6).

Another group of sulphur-oxidising bacteria characterised by members of the green *Chlorobium* and purple *Chromatium* genera have the capacity to remove hydrogen sulphide. These bacteria of mud and stagnant water possess the ability to carry out photosynthesis in anaerobic conditions. Instead of splitting water like higher plants and using this process ('Photolysis') to donate electrons to their photosystems, they split hydrogen sulphide and release elemental sulphur by a process known as 'photoautotrophy'. Electrons released from the hydrogen sulphide are then pushed by the action of light upon the photosystems to potentials high enough (even more electronegative than the standard hydrogen potential at 25 °C and pH 7) to allow electrons to flow downwards through certain biological couples (iron–sulphur proteins called 'ferredoxins') to reduce a vital biosynthetic oxidation–reduction cofactor called nicotinamide adenine dinucleotide ($NAD^+/NADH$) as shown in Fig. 2.5. The protons released from the hydrogen sulphide then cause the generation of a gradient of protons (or ΔpH) across membranes and this may be converted into more chemical energy in the form of ATP by a chemiosmotic process (see Appendix, Sect 8.3). These photoautotrophic bacteria have considerable evolutionary significance (being earlier responsible for the formation of the major proportion of the geological

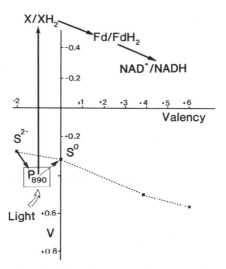

Fig. 2.5 Electron flow during photoautotrophy in sulphur- oxidising photosynthetic bacterial genera such as *Chromatium* or *Chlorobium*. Note that the energy from the light-trapping process associated with the photosystem (P_{890}) is used to form highly electronegative reduced compounds (XH_2) the structure of which are not yet known, which then reduce iron–sulphur proteins (FdH_2) known as 'ferredoxins'. These in turn reduce important biological cofactors such as nicotinamide adenine dinucleotide (NADH).

sulphur reserves of the world) but they have little importance to the global cycling of sulphur today.

More unpleasant anaerobic organisms use the amino-acids hydrolysed from the proteins of decaying bacteria, plants and animals as a source of carbon compounds from which they derive energy. As they do so, the sulphur and nitrogen of the amino-acids are released as hydrogen sulphide and ammonia without any change in oxidation or reduction state (see Fig. 2.6). These products of a process known as 'desulphurylation' contribute greatly to the stench of putrefaction – a nasty but necessary component of the cycling of sulphur.

A number of microbial processes are also known which cause the reduction of oxidised forms of sulphur. To do this, electrons and energy have to be supplied from elsewhere to convert more electropositive oxidized forms of sulphur into more electronegative reduced compounds. These may come from the normal photosynthetic mechanisms of higher plants and algae which then reduce sulphate to form the sulphydryl groups (—SH) of sulphur-containing amino-acids and other compounds by a process known as 'assimilatory photosynthetic

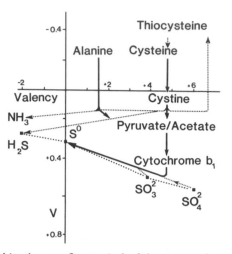

Fig. 2.6 Anaerobic electron flow typical of the process known as 'desulphurylation' which is a major component of putrefaction. Note that, in this case, the reduced state of sulphur (or nitrogen) from the amino-acids does not change as they are released but sulphur respiration (see Fig. 2.7) may also be present.

sulphur reduction' (see later). Alternatively, a trapped food source such as glucose may be respired aerobically to generate the necessary energy and electrons by a process known as 'assimilatory hetero-trophic sulphur reduction'. Most fungi, yeasts and aerobic bacteria, for example, do this in order to synthesise the sulphur-containing amino-acids they need. Consequently, these two types of assimilatory sulphur reduction are the major routes by which highly oxidised sulphate is adsorbed and trapped by the biosphere from atmospheric sources. Inevitably, more sulphur is trapped than is needed and this causes volatilization of the excess (mainly as hydrogen sulphide, see Ch. 6) but in these aerobic organisms, unlike more unpleasant anaer-obes, the uptake of sulphur usually balances that required for nu-tritional purposes and only a small proportion of sulphide is released or accumulated.

The same cannot be claimed for another anaerobic process known as 'dissimilatory' or 'respiratory sulphur reduction' which is carried out by bacteria in the genera *Desulfovibrio* and *Desulfotomaculum*. These organisms, found in water-logged soils, mud, stagnant ponds, sewage pipes and the rumens of grazing animals, take in large amounts of sulphate or sulphite and reduce them to elemental sulphur or to sulphides. These products are then released in the immediate vicinity or stored, only a small fraction of the latter going towards synthesis

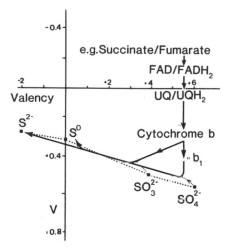

Fig. 2.7 Electron flow typical of respiratory sulphur reduction (often called 'dissimilatory sulphur reduction') as it occurs in the anaerobic bacteria in the genera *Desulphovibrio* or *Desulfotomaculum* through biological couples which include reduction of the flavoproteins ($FADH_2$) and ubiquinone (UQH_2).

of sulphur-containing amino-acids. Usually dissimilatory 'sulphur reduction' uses a reduced carbon source as a source of electrons and energy is trapped as ATP whilst they flow through biological couples. As sulphate is used as a terminal electron acceptor in the absence of oxygen (see Fig. 2.7) this process is often called 'sulphur respiration'.

By comparison to the assimilatory sulphur reduction of plants, fungi and aerobic bacteria (see later), sulphur respiration probably plays only a small part in removing sulphur from the atmosphere because of the restrictions placed upon these organisms by having to live in an anaerobic environment. However, they do cause environmental problems, not just due to the smells they cause when their habitat is disturbed, but because they are a major agent in causing the corrosion of underground metal pipework. Normally, when iron pipes are in contact with water, some of the metal dissolves and hydrogen gas is formed (Reaction 2.12). If this is not removed an opposing electro-chemical potential is formed (i.e. it becomes polarised) and the iron stops dissolving. Some dissimilatory bacteria, however, possess the ability to utilise hydrogen, as a source of electrons instead of a carbon source to reduce the sulphate (Reaction 2.13). This depolarises the system and permits accelerated corrosion to take place (Reaction 2.14).

$$Fe + 2H_2O \longrightarrow Fe^{2+} + 2OH^- + H_2 \qquad (2.12)$$

$$SO_4^{2-} + 4H_2 \longrightarrow S^{2-} + 4H_2O \qquad (2.13)$$

$$S^{2-} + 2Fe^{2+} + 4OH \longrightarrow FeS + Fe(OH_2) + 2OH^- \quad (2.14)$$

(d) Reactions of sulphur dioxide in water and tissue fluids

Sulphur dioxide molecules contain no unpaired electrons and thus are not free radicals. They readily dissolve in water to form sulphite (SO_3^{2-}) and bisulphite (HSO_3^-) ions in a series of reactions (Reaction 2.15). Un-ionised sulphur dioxide in solution may be considered either as sulphurous acid (H_2SO_3) or a dissolved form of sulphur dioxide gas ($SO_2.H_2O$) but the equilibrium constants governing the exchanges between sulphurous acid and bisulphite are so much in favour of the latter that very little unionised dissolved sulphur dioxide exists in solution anyway. The relative proportions of bisulphite and sulphite in solution are more important because the dissociation equilibrium (scc Appendix Sect 8.5) between the two has a pK_a value of 7.18 which means that around neutrality (commonly regarded as a pH approximating to conditions prevailing in biological tissues) roughly equal proportions exist (see Fig. 2.8).

$$SO_2 + H_2O \rightleftharpoons H_2SO_3^- \rightleftharpoons H^+ + HSO_3^- \rightleftharpoons H^+ + SO_3^{2-} \quad (2.15)$$

Both sulphite and bisulphite have a lone pair of electrons on the sulphur atom which means that attack on electron-deficient sites in other molecules is strongly favoured. Furthermore, bisulphite or sulphite may be readily oxidised by a series of overlapping reactions which involve the formation or consumption of free radicals (Reactions 2.16 to 2.21; note, see Appendix, Sect. 8.2, for an explanation of the nature of free radicals). The presence of a metal such as manganese appears to be vital for some of these rapid oxidations (Reactions 2.16 and 2.17) and the production of superoxide ($O_2^{\bullet-}$) by two different reactions (Reactions 2.16 and 2.18) is confirmed by sensitivity to superoxide dismutase (see Ch. 5) only when chelating EDTA is present to mop up metals like manganese thus precluding a route from Reaction 2.17 directly to Reaction 2.19 which does not generate superoxide free radicals.

$$HSO_3^- + O_2 \xrightarrow{Mn^{2+}} HSO_3^{\bullet} + O_2^{\bullet-} \qquad (2.16)$$

$$SO_3^{2-} + O_2 + 3H^+ \xrightarrow{Mn^{2+}} HSO_3^{\bullet} + 2OH^{\bullet} \qquad (2.17)$$

$$HSO_3^{\bullet} + O_2 \longrightarrow SO_3 + O_2^{\bullet-} + H^+ \qquad (2.18)$$

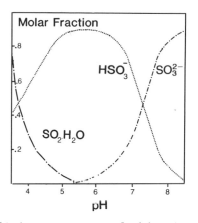

Fig. 2.8 Relationship between amounts of sulphite, bisulphite and dissolved, but undissociated, sulphur dioxide ($SO_2.H_2O$) at different pH values.

$$HSO_3^{\cdot} + OH^{\cdot} \longrightarrow SO_3 + H_2O \qquad (2.19)$$

$$2HSO_3^{\cdot} \longrightarrow SO_3 + SO_3^{2-} + 2H^+ \qquad (2.20)$$

$$SO_3 + H_2O \longrightarrow SO_4^{2-} + 2H^+ \qquad (2.21)$$

Other free radicals of sulphur have also been detected in addition to the bisulphite ($HSO_3^{\cdot}$) species implied by Reactions 2.16 to 2.20. For example, the action of light or chemical and biological reducing agents are capable of forming sulphoxyl radicals ($SO_2^{\cdot-}$) which persist for much longer than most free radicals. These may be the reactive agents that donate electrons to electron-deficient sites as well as allowing another means of oxidising sulphite to sulphate (Reactions 2.22 and 2.23).

$$SO_2 + OH^- \xrightarrow[\text{or reducing agents}]{\text{light}} SO_2^{\cdot-} + OH^{\cdot} \qquad (2.22)$$

$$SO_2^{\cdot-} + O_2 + H^+ + e \longrightarrow HSO_4^- \qquad (2.23)$$

It is always desirable that levels of uncontrolled free radicals should not rise within living tissues because so many other troublesome reactions may take place. As both Chs. 3 and 5 emphasise, most tissues have comprehensive free-radical scavenging systems which mop up free radicals as soon as they are formed. The case of sulphite is no exception. Rather than rely upon the above reactions to oxidise sulphite to sulphate, nearly all organisms have a molybdenum-containing enzyme called sulphite oxidase which ensures rapid removal of sulphite. In higher plants, fungi and animals, this enzyme is located in the energy-forming mitochondria where electrons normally flow from reduced organic acids to oxygen forming water (see Appendix, Sect. 8.4). Some electrons, however, are diverted to sulphite via cytochrome *c* to oxidise sulphite to sulphate. The essential, free-radical-avoiding nature of this enzyme is well illustrated by a congenital disease which occurs in few babies who lack sulphite oxidase. As a consequence, this deficiency causes such severe neurological disturbance that the infant's brain tissues are extensively damaged by free-radical action leading to death.

(e) Reactions of sulphite with biomolecules

The number of possible interactions of sulphite or its products with other critical biochemical compounds is very large. It is quite evident from Reactions 2.16 to 2.23 that a wide variety of free radicals ($HSO_3^{\bullet}$, $O_2^{\bullet -}$, $OH^{\bullet}$, $SO_2^{\bullet -}$) may be formed as sulphite is oxidised and these would have a wide range of consequences. The 3 free-radical possibilities which have been considered highly likely in the presence of oxygen and manganese are chain cleavage of DNA, the oxidation of the double bonds in fatty acids within membranes and the conversion of the amino-acid methionine into methionine sulphoxide (see Chapter 5).

However, bisulphite and sulphite are both highly reactive nucleophiles in their own right. Indeed, the efficiency of some of their reactions with organic compounds forms the basis of the widely used practice of food preservation by sulphur dioxide. Amateur wine makers will also be very familiar with the sterilising capabilities of sodium sulphite tablets. In high concentrations, sulphite is also a highly specific mutagenic agent by virtue of the conversion of cytidine to uracil (Fig. 2.9) within DNA causing guanine–cytidine base pairs to turn into adenine–thymidine associations because uracil reads the same as thymidine upon replication of nucleic acids. The sterilising effect of sulphur dioxide, however, resides not just in specific reactions like this but in causing many other disruptive reactions within microbes, killing them immediately. Chief amongst these is the attack on disulphide bridges within enzymes and structural proteins (Reaction 2.24, where R and R′ represent distinct peptide components of a protein). This inactivates critical reactive centres of hydrolytic enzymes in microbes, or the food itself, thus retarding breakdown and undesirable microbial contamination.

$$R–S–S–R' + HSO_3^- \longrightarrow R–S–SO_3^- + R'SH \qquad (2.24)$$

Other important nucleophilic attacks take place on important

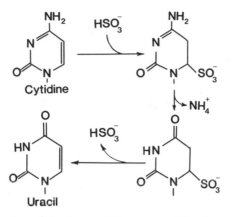

Fig. 2.9 Conversion of the base cytidine to uracil by the action of bisulphite may be responsible for some of the mutagenic properties of such anions acting upon bases within nucleic acids.

biological ring compounds such as NAD⁺ (see Fig. 2.10), NADP¹, FAD, FMN, pteridines, folate, tryptophane and thiamine. The latter is a particular problem for the food industry because the attack of sulphite on thiamine (Fig. 2.10), a vitamin of the B complex, is so efficient that some sulphur dioxide preserved foods have to be fortified with thiamine by manufacturers to offset the possibility of thiamine deficiency arising.

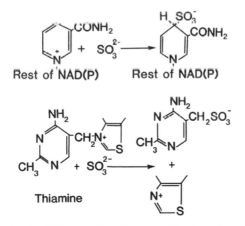

Fig. 2.10 Attack of sulphite on the nicotinamide ring of pyridine-linked dehydrogenases or destruction of the vitamin thiamine are likely events.

However, during exposure to atmospheric sulphur dioxide, the levels of the reactants such as sulphite that arise are much too low to allow significant interactions similar to those that occur during food preservation. In the latter, one of the reactants (sulphur dioxide) is in vast excess. A reaction that has often been put forward as one capable of producing low levels of highly toxic products is the formation of α-hydroxysulphonates (Reaction 2.25). Such compounds are known to interfere, for example, with various reactions of photo-respiration (see Sect. 6.1c) in plants involving glycollate (HOCH.COO⁻) at very low concentrations. Furthermore, the sulphonated adducts formed by attack of sulphite on ring systems of NAD^+ (Fig. 2.10), $NADP^+$, FMN and FAD may be stabilised when these cofactors are bound to enzymes like lactate dehydrogenase – which then prevents any significant back-reaction. This means that significant amounts of a number of enzymes may be rendered inoper-ative for some time. During exposure to atmospheric sulphur dioxide the inhibitory consequences, like these involving cofactors of critical enzymes disabled in this manner or those specifically inhibited by α-hydroxysulphonates, could be considerable.

$$R.CHO + HSO_3^- \longrightarrow R.CH.OH.SO_3^- \qquad (2.25)$$

2.2   PLANTS

(a)   Stomatal access

The first hindrance encountered by a gaseous pollutant before reaching a leaf is a layer of relatively still air above and below the surface of the leaf. This may constitute a considerable aerody-namic or boundary layer barrier to the entry of atmospheric pollutants. Unstirred air layers on both sides of a leaf thus represent an important resistance (see Fig. 2.11) to the access of sulphur dioxide. Movement across such an obstruction is therefore achieved by dif-fusion of gas molecules in response to differences in concentration. The extent of this boundary layer resistance varies with both wind speed and various leaf properties such as size, shape and orientation. As wind speed increases, resistance to the inward movement of pollutant molecules falls, uptake increases and in this way increased wind speed can facilitate uptake of pollutant. The boundary layer is generally thinner at the edges of a leaf than at the centre which might account for increased pollutant damage often seen at margins of leaves, especially on grasses and cereals.

Although epidermal cells occupy a much greater proportion of a leaf surface than do stomatal pores, the waxy cuticle covering them

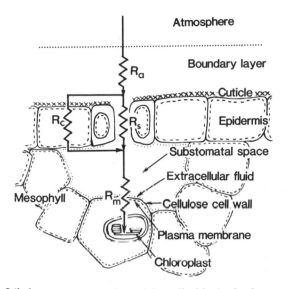

Fig. 2.11 Likely access routes for sulphur dioxide (and other gaseous air pollutants) into a plant leaf. The layer of still air or boundary layer imposes a resistance to flow, R_a, which depends on a number of factors including wind speed. Access then may be restricted through the degree of opening of the stomata, R_s, or by penetration of the cuticle and epidermal layers, R_c. In the case of sulphur dioxide and ozone, entry through the stomata is thought to be the major route of access but in the case of the oxides of nitrogen the situation may be reversed (see Ch. 3). The mesophyll resistance, R_m, consists of a number of different components: the still air of the substomatal space, the extracellular fluid, the cellulose cell wall, the plasma membrane, the cytoplasm, and any organelle envelope membranes before the major site of reaction is reached

offers a greater barrier to penetration by most pollutant gases. As a consequence, this cuticular resistance (Fig. 2.11) is generally very large and much greater than that through the stomatal pores and into the internal air spaces beneath. Cuticular resistance is often ignored in any consideration of pollutant gases entering a leaf. Nevertheless, sulphur dioxide deposited on leaf and stem surfaces may dissociate in any water film and react with cuticular waxy components to damage cuticular surfaces. In these circumstances, a certain amount of the gas may enter leaves by penetrating the damaged cuticle although most it still enters through the stomata. Alternatively, the sulphur dioxide may stimulate the production of protective waxes in certain species which in time increases the cuticular resistance.

When the cuticle is wet for long periods, especially at night or in the winter, the apparent deposition rate of sulphur dioxide on to a leaf surface may be considerably enhanced despite the fact that the stomata are closed. This has additional consequences especially for conifers which are considered later in Chs. 4 and 7.

For stomata to open, the guard cells have to be turgid which causes the gap between them to widen. In most plants this occurs during the day as normal exchanges of water and carbon dioxide take place. Thus uptake of gases like sulphur dioxide from the atmosphere into plants also takes place mainly during the day. In very dull conditions, stomatal opening will not be extensive, so rates of uptake will be reduced. At night when stomata are closed, rates of removal of sulphur dioxide or ozone by plants will approach those of other inert surfaces. Consequently, those changes in stomatal aperture which provide a plant with control mechanisms for adjusting the movement of carbon dioxide into leaves and water vapour loss at the same time also influence pollutant gas uptake. Stomatal behaviour, frequency and distribution are therefore important factors in affecting the amount of pollutant a particular plant receives.

At high humidities, guard cells respond to the presence of sulphur dioxide in a manner different to almost all the other compounds which are known to have an effect upon stomata. Sulphur dioxide causes them to become even more turgid and therefore the pores open wider still, allowing more polluted air to enter the leaf than would normally be the case. If the pollutant is removed quickly the effect is reversible (although less so if prolonged) but it does mean that in wet conditions the access of sulphur dioxide and any other accompanying pollutant is made easier. The converse is true in dry conditions when humidities are low. Stomatal pores are open less and hence the resistance to pollutant entry is greater but this closure may be due to other drought-induced protective mechanisms inherent to a plant rather than a direct effect of sulphur dioxide similar to that caused by wet conditions. The opening response in high humidities due to sulphur dioxide may have long-term effects under natural conditions, especially in conifers if any increase in water loss occurs. The mechanism of action of sulphur dioxide which causes these turgor changes in guard cells has not been fully investigated but it is presumed that pH changes and enhanced levels of bisulphite and sulphite disturb the fluxes of potassium, calcium, chloride, malate and protons between guard cells and their subsidiary cells which are normally involved in stomatal response. My colleagues at the University of Lancaster have recently shown that calcium-regulated channels through the guard cell membranes are part of certain closing responses. It is possible that sulphite (or bisulphite) in some way interferes with a calcium-dependent mechanism.

(b)   Internal resistance and buffering

Just as access of sulphur dioxide across the boundary layer and through the stomatal pores can be treated as variable resistance (Fig. 2.11), the movement of the polluting gas – and its products sulphite or bisulphite – to their sites of action across the substomatal cavity in the extracellular fluid on the internal mesophyll cell walls can be similarly treated. This mesophyll resistance (R_m) is neither constant nor so small as to be negligible. It is likely to vary between species and even between different varieties. Certain aspects of pollution tolerance or pollution sensitivity to sulphur dioxide within a species may even be partly due to differences in mesophyll resistance. Moreover the mesophyll resistance is not a simple diffusive resistance because the sulphur dioxide and its allies have to cross the cellulose cell wall, the cell (or plasma) membrane, and some of the inner cytoplasm and even double organelle envelope membranes before they reach a likely target (the chloroplast, see later). As they do so they are likely to disrupt the membranes, change the ionic environment of the cytoplasm and cause a number of other secondary changes which in turn affect the route through which they pass.

The behaviour of sulphur dioxide in water and tissues has already been covered (Sect. 2.1d) but one of the most critical determinants will be the buffering capacities in the extracellular fluid, the cytoplasm and inside any target organelle such as the chloroplast. Buffering (i.e. the ability to counteract a change of pH from a median value) in cells is carried out by means of a wide variety of molecules both large (often proteins) or small (e.g. amino-acids), with free carboxyl, phosphate, amino or sulphydryl (—SH) groups. These may be momentarily overloaded if high doses of sulphur dioxide and its allies are suddenly experienced for a short time whereas the same dose at low concentration over a longer period would present no problem. It has been claimed that the leaves of plants exposed to sulphur dioxide have a lower buffer capacity than the equivalent tissues of unexposed plants. If pH should change in certain parts of the cell (e.g. the chloroplast) the consequences upon cellular metabolism could be significant.

As discussed earlier (Sect. 2.1d), a variety of different dissociation equilibria exist for the products of sulphur dioxide in solution. Normally, un-ionised weak acids easily enter cells and organelles as they may freely cross membranes. On the other hand, fully dissociated strong acids produce anions which normally do not passively cross membranes. However, certain embedded proteins in membranes called 'pumps' or 'translocators' actively pass charged species from one side of a membrane to another. The phosphate translocator of the inner chloroplast envelope is one such example. This transport protein spanning a membrane behaves as an antiporter because it exchanges inorganic or organic phosphate going one way for a different organic form of phosphate coming the other way. Moreover, it also allows sulphite or sulphate to travel across the membrane in a 'piggy-back' fashion. Thus strong acids like sulphurous or sulphuric acids gain access to cellular compartments more readily than passive diffusion would normally allow.

Algal studies have shown that there is generally a positive linear relationship between cytoplasmic pH and external pH but in more advanced plants the ability to resist and control intracellular gradients of acidity and alkalinity is much greater. In lichens, however, control of pH by a 'pH-stat' mechanism is very primitive and relies extensively, but not entirely, on the inherent buffering capacity. It is commonly believed that lichens as a whole are especially sensitive to sulphur dioxide atmospheric pollution because of their scarcity in heavily polluted areas. This is true for only some lichens. More importantly, regions in Northern Europe and the United States also have high humidities which normally encourage luxuriant growth of lichens so the likelihood of damage is that much greater. In addition, there is a greater chance of other biotic and abiotic factors within these areas which may also limit the growth of certain lichens. Therefore, taken as a group, lichens are no more sensitive to sulphur dioxide pollution than higher plants although they have proved most useful as biological indicators of the long-term presence of adverse atmospheric pollution conditions within a particular locality or region.

In higher plants, the mechanism of the pH-stat is complicated and only partly understood. Buffering capacity plays a part but proton pumps exist in membranes which allow protons to be transferred from one compartment to another within the cells or to the outside. One most interesting observation made recently has shown that the proton pump of the cell (or plasma) membrane is unaffected by either sulphate or nitrate (which may arise from nitrogen dioxide pollution, see Ch. 3) but the tonoplast (the membrane bounding the large fluid-filled space or vacuole of plant cells which is often filled with acidic storage or waste materials) proton pump is strongly inhibited by nitrate and, to a lesser extent, by sulphate. In other words, if anions like sulphate or nitrate (and probably sulphite, etc.) gain access to the cytoplasm from extracellular fluid on the outside, then the normal pH-stat mechanism of regulation by increasing export of protons to the acidic vacuole is hindered and the only effective means of pH control is to pump harder outwards against a concentration gradient already made worse by atmospheric pollution coming in. This means more energy is needed to resist pH changes and therefore less is available for growth.

(c)   Sulphur metabolism

As already mentioned in Sect. 2.1(c) (in which the microbial exchanges of sulphur were discussed) the higher plants and the algae are capable of using their photosynthetic electron flow to reduce sulphate to sulphydryl groups (-SH) needed for the synthesis of sulphur-containing amino-acids by a process known as 'assimilatory photosynthetic sulphur reduction'. The reduction of sulphate appears to have two pathways (Fig. 2.12); one which involves the 'active' sulphur cycle and one which is analogous to the pathway of nitrate reduction in plants (see Ch. 3). Sulphur reduction involves the transient formation of sulphite from sulphate which then is further reduced to sulphide. The situation is complicated because sulphite is also rapidly photo-oxidised to sulphate in the chloroplast as well as being oxidised by sulphite oxidase elsewhere in the mitochondria. It would appear that the superoxide free radical ($O_2^{\cdot}$) is implicated in some way in the photo-oxidation although low levels of sulphite can donate electrons to Photosystem II and the $HSO_3^{\cdot}$ free radical which is formed then accepts electrons from Photosystem I to become sulphite again.

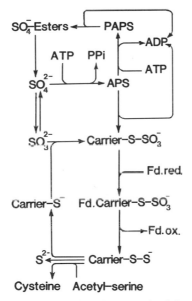

Fig. 2.12   Biological oxidation and reduction of sulphate and sulphite. Uppermost is the so-called 'active' sulphur cycle involving the sulphonated nucleotides, adenosine-5'-phosphosulphate (APS) and adenosine-3-phospho-5'-phosphosulphate (PAPS). The sulphate on former may also be reduced to sulphide by a series of reactions (lower half) involving the iron–sulphur protein, ferredoxin (Fd).

This means that natural low levels of sulphite exist in close proximity to the membranes of a chloroplast which are repeatedly recycled by this futile oxidative process and at the same time the possibility of reduction to sulphydryl groups may take place. Only when the input of sulphite from sulphur dioxide fumigation becomes excessive does this delicate balance fail and problems arise from enhanced free-radical activity which are not held in check by scavenging processes.

The reduction of sulphite in chloroplasts produces sulphide, especially hydrogen sulphide (H_2S, see Ch. 6) as well as sulphydryl groups. The release of hydrogen sulphide by leaves in the light but not in the dark would appear to be a route by which plants may rid themselves of excess sulphur caused by exposure to sulphur dioxide, although few experiments have been done to establish this and some believe that this detoxification does not occur. The problem resides in the way many of these experiments are carried out. It is a frequent biological practice that, in order to elucidate the fate of a compound, a tag is added which can be followed. In the case of sulphur, the radioactivity associated with ^{35}S is often used. To be certain of establishing the pathways into all the possible exits, a high concentration of $^{35}SO_2$ is used over a short period for more convenient experimentation. However, this does not mimic the actual exposure conditions of plants to low doses of sulphur dioxide. When these objections have been covered by more careful attention to the exposure conditions, most of the radioactivity associated with ^{35}S is found to remain in the plant as sulphate, sulpholipid, sulphur amino-acids or other sulphur-containing materials.

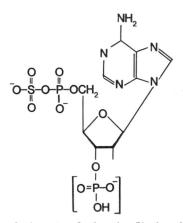

Fig. 2.13 Structure of adenosine-3-phospho-5′-phosphosulphate (PAPS). If the lower phosphate (in square brackets) is removed and replaced by hydrogen then the structure is that of adenosine-5′-phosophosulphate (APS).

Some of the sulphate is taken up either by the 'active' sulphur cycle (Fig. 2.12) which involves the formation of two so-called 'high-energy'compounds related to ATP; adenosine-5'-phosphosulphate (APS) and adenosine-3-phospho-5'-phosphosulphate (PAPS, see Fig. 2.13) which are involved in the sulphation of organic molecules (mainly lipids). Alternatively, sulphur reduction produces sulphur-containing amino-acids such as cysteine (Fig. 2.12). Once formed, this amino-acid is the basis for a series of interconversions which allow the formation of all the other sulphur-containing amino-acids (Fig. 2.14). The biosynthetic sequence from cysteine to methionine (solid lines) is basically the same in plants, fungi and most microbes and little breakdown occurs. The situation is reversed in man. Most monogas-

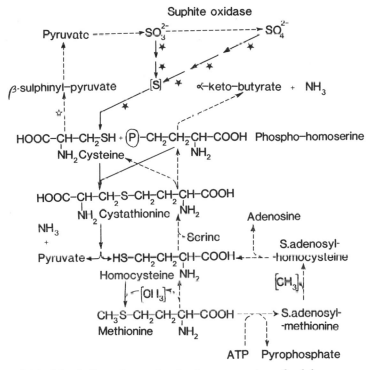

Fig. 2.14 Metabolic pathways for the interconversion of sulphur compounds (especially sulphur-containing amino-acids) in biological tissues. Those shown with solid arrows are mainly employed in plant tissues whereas the breakdown of sulphur amino-acids (dashed arrows) are mainly associated with animals. Reactions beyond the open asterisk are normally associated with mitochondria and those marked with a solid asterisk are mainly those already shown in Fig. 2.11.

47

tric animals require sulphur-containing methionine in their diets because they lack most of the biosynthetic enzymes between inorganic sulphur and either cysteine or methionine. Therefore, in such animals, emphasis is placed upon metabolic breakdown (dashed lines, Fig. 2.14) which gives rise to cysteine, β-sulphinyl pyruvate, taurine, pyruvate, α-ketobutyrate, ammonia and sulphite. The latter again requires immediate oxidation to sulphate by sulphite oxidase before most of them are eliminated through the kidneys (see Sect. 2.3c).

Ruminant animals differ from man and other monogastric animals because they can use inorganic sulphur from herbage as their only source of sulphur. The reason is that the microbes of the rumen (which precedes the true stomach – the abomasum) either reduce sulphate to sulphide, or remove sulphur from sulphur-containing amino-acids by desulphurylation (see Sect. 2.1c) and then resynthesise their own sulphur amino acids from sulphide. The bacterial proteins and amino-acids are then released when digestion takes place in the abomasum at pH 1 and are subsequently absorbed by the ruminant through the intestinal walls. This means that grazing animals feeding on herbage which is deficient in sulphur-containing amino-acids need only have inorganic sulphate added to their diets. In practice it is often claimed that supplementation is never needed because, even on soils which would be naturally low in sulphur, atmospheric deposition of sulphur dioxide is usually more than adequate to compensate for this deficiency.

(d) Chloroplastidic damage

Inhibition of photosynthesis is frequently thought to be one of the first effects of sulphur dioxide upon plants and therefore the chloroplast is often regarded as the primary site of many of the disturbances caused by sulphur dioxide or its products in aqueous solution. The pH of the chloroplast stroma is generally much greater than pH 7 (nearer pH 9 in the light) and this favours the formation of sulphite ions at the expense of bisulphite when sulphur ionises in solution (Fig. 2.8). As a consequence, the effects of sulphite are often considered to be a reflection of the action of sulphur dioxide within chloroplasts but, if the pH is lowered, sulphur compounds will enter more readily as dissolved $SO_2.H_2O$ (Fig. 2.8) and all forms of sulphur dioxide will become better oxidising agents.

Ultrastructural studies of the effect of sulphur dioxide fumigation upon plants have been undertaken. These have shown a swelling of the lumen spaces within the thylakoids (see Appendix, Sect. 8.3) is one of the first effects of sulphur dioxide upon plants. Initially, this swelling is a reversible phenomenon although the time taken for recovery is proportional to the dosage. Such swelling is indicative of

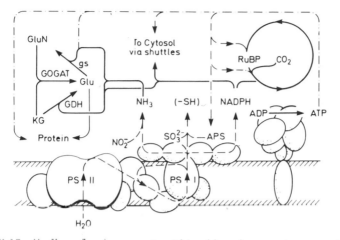

Fig. 2.15 Outline of various events within chloroplast membranes and stroma. The flow of electrons through the photosystems are used for a variety of different purposes (i.e. reduction of $NADP^+$, nitrite, sulphite or APS) but they also induce a pH gradient across the membrane (not shown) which is harnessed by the coupling factor particles to form ATP (lower right, see Appendix, Sect. 8.3, for details). The uses of ATP are many but a number of the most important are shown in this diagram, all of which may be detrimentally affected by sulphur dioxide and its allies. (Courtesy of Butterworth Sci. Press, London.)

the ionic disturbances and the acidification mentioned earlier (Sect. 2.2b).

Figure 2.15 summarises in an idealised manner some of the more important features of photosynthetic events as they occur within chloroplasts. This includes the biochemistry of not just the fixation of carbon dioxide but also implicates sulphur and nitrogen metabolism. All such events are dependent upon electron flow from water induced by the photosystems in the presence of light. Electron transport also creates a proton gradient across the thylakoid membrane causing the stroma to become more alkaline and the lumen more acidic. This alkalinisation enhances the activity of enzymes within the Calvin cycle and any decrease in pH due to additional ions from sulphur dioxide pollution are therefore likely to reduce the efficiency of carbon dioxide fixation. The proton gradient or $\triangle$pH across the membrane is utilized by the chloroplast coupling factor complex ($CF_0 + CF_1$) to form ATP (see Appendix, Sect. 8.3, for fuller description). Figure 2.15 emphasises only a few of the many biosynthetic demands placed upon the availability of ATP. Any detrimental effect of sulphur dioxide or its allies upon any of the processes leading to photophosphorylation (i.e. the formation of ATP from ADP and ortho-phosphate by light) such as impaired operation of the photosystems, interference with electron flow, leakiness of the membrane to protons, failure to utilise the proton gradient, or inhibition of the site of phos-phorylation upon chloroplast coupling factor (CF_1) particles will all cause a reduction of rates of many biosynthetic activities, particularly protein or starch synthesis and carbon dioxide fixation.

Direct effects of sulphur dioxide and/or sulphite upon the photo-systems were originally thought to be prime causes of reductions in net photosynthesis but only unrealistically high levels of sulphur dioxide and its allies have actually been shown to cause inhibition of electron flow through the photosystems. There is more support for direct effects of sulphur dioxide (at realistic levels) upon both carbon fixation in the stroma and the photophosphorylation of ATP by the coupling factors (right-hand side of Fig. 2.15). The latter is especially noticeable at low phosphate levels and may be due to sulphite competing for phosphate sites during ATP formation. This in turn may lead to a shortage of ATP which is needed for fixation of carbon dioxide, for export to the rest of the cell or for the formation of protein (Fig. 2.15). There is good support for this 'ATP deprivation' concept because reductions in ATP pool sizes in pines, for example, correlate with increases in atmospheric sulphur dioxide levels. Similar reductions in ability to form ATP have been detected in plant mitochondria.

Research time has been also devoted to the study of direct effect of sulphur dioxide and its allies upon enzymes in the chloroplast stroma, particularly those involved in the fixation of carbon dioxide. Positive evidence of inhibition has been established especially upon those enzymes that are regulated either directly by light or those that require a light-dependent pH change before activation. The critical enzyme of carbon dioxide fixation, ribulose-1,5-*bis*-phosphate carboxylase/oxygenase (given the 'breakfast cereal manufacturer' acronym of RubisCO), for example, requires a pH change in the stroma from pH 7.5 to pH 9 before becoming active. The question that still remains is: 'Does an enzyme like this respond directly to sulphite and a portion is prevented from becoming active because the reactive site is blocked, or does the general acidification and pressure upon buffering by sulphur dioxide and its allies then reflect on reduced activation?' This question is still not resolved and probably the answer lies in a combination of the two possibilities. It has been calculated that a reduction of chloroplast stroma by 0.5 of a pH unit would cause a 50 % decrease in net photosynthesis so even a small change could be significant to the growth of a plant in polluted conditions over a long period.

The integrity of the photosynthetic membranes to maintain a proton gradient across the thylakoid membrane (which is generated by electron flow and harnessed by the coupling factors to make ATP, see Appendix, Sect. 8.3) is also disturbed by the presence of sulphite, etc., especially in the presence of other anions like nitrite. It appears that both sulphite and nitrite together are capable of promoting additional free-radical formation which then causes the membranes to become 'leaky' with respect to protons. If this should occur then insufficient ATP would be produced and reduced growth would be the final outcome. This possibility is discussed further in Ch. 7.

(e) Long-term effects

There are hundreds of publications in which exposure to sulphur dioxide is claimed to exert an effect (usually adverse) upon crops, trees, lichens and mosses. Unfortunately, it is extremely difficult to establish valid quantitative conclusions from this mass of literature because the results are so variable. This is because various parts of different species, cultivars or clones of plants are studied at different seasons in different environmental conditions (soil, temperature, humidity, etc.) with a wide variety of exposures to pollutants using open-top chambers, fumigated field plots, controlled environment cabinets, wind tunnels and comparisons between different localities. In earlier days research naturally focused on obvious visible injury (i.e. chlorosis, necrosis, leaf fall and dead leaves, etc.), but it is quite clear that sulphur dioxide may also reduce the growth and yield in the absence of visible symptoms which can only be evaluated by having a valid control alongside for comparison. This raises a fundamental question, 'what is a valid control?'

In the past some have argued that a valid control is one in which plants are grown in atmospheres containing no pollutant at all (i.e. charcoal-filtered air). The problem here is that this situation is almost as unrealistic as some of those long-term exposures to continuous high levels of sulphur dioxide which have actually only occurred momentarily as peak values in the real world. Comparisons of plant growth in charcoal-filtered air against even those obtained with low levels of sulphur-dioxide-polluted atmospheres are remarkable. Growth differences of as much as 30 % are often measured – but 30 % of nothing is still nothing. In many instances, the charcoal-filtered air situation is an idealised situation and the low-level treatment could in fact represent a much better and realistic 'clean-air' control upon which comparisons may be then made. Comparative work of this kind using clean-air controls can therefore only be done in areas known to have little sulphur dioxide pollution and the results from charcoal-filtered experiments carried out in areas with high sulphur dioxide background pollution treated with some caution.

They are measuring quite different differences. Moreover, filters such as activated charcoal, however, also remove nitrogen dioxide, ozone, PAN or hydrocarbons and therefore hidden interactions between pollutants may have been unwittingly exposed. Consequently both types of experimental controls may have their place. Although never actually done, the comparative experiments should have two controls, one clean-air and one charcoal-filtered. A further word of caution, activated charcoal does not remove nitric oxide so even charcoal-filtered air could still be contaminated.

From studies so far it has been concluded by some researchers that the sulphur dioxide concentrations prevailing in most agricultural regions of Europe, Northern America and elsewhere are not high enough significantly to reduce the yield of any of the major cereals. However, concentrations of 60 ppbv sulphur dioxide are capable of causing at least a 7.5 % reduction in one of the commonest grasses found in permanent pasture (perennial ryegrass, *Lolium perenne* L.) and according to an OECD cost-benefit analysis in 1981 this may be as high as a 25 % reduction. Yield reductions in grasses are probably more closely related to dosage (concentration × time) but are complicated by the additional presence of other pollutants especially nitrogen dioxide which cause more-than-additive (sometimes called synergistic) reductions in growth (see Ch. 7).

A wide variety of environmental conditions affect the growth of plants particularly grasses and trees. As mentioned before, the rate of air movement over a sward or canopy can influence sensitivity to sulphur dioxide. If the wind speed is high enough, the boundary layer resistance (see Sect. 2.2a) is significantly reduced as the layer of still air directly above the leaves or needles is ripped away. Seasonal factors also need to be taken into account. Generally it has been found that, in grasses at least, the inhibitory effects of sulphur dioxide are greater (by as much as 50 % in the late winter) under conditions of slow growth caused by low light levels (with and without short day lengths) and by low temperatures.

Stress, which the purist would call 'distress', is an additional imposition upon plants. It may also cause variations in response when associated with the additional presence of sulphur dioxide. Figure 2.16 is a conceptual model built up by the contributions of a number of UK researchers which emphasises the point that pollution injury cannot be divorced from other damaging conditions; each one is interrelated. Furthermore, age and the spacing of plants are also important determinants. With all these factors to be considered,it is not surprising that such divergent and apparently contradictory results concerning the effect of sulphur dioxide on plants have been reported in the past.

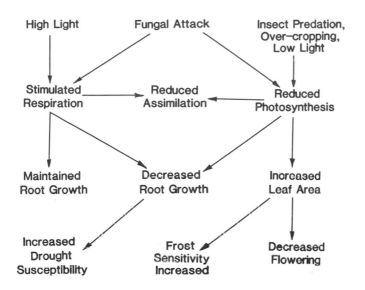

Fig. 2.16 The interrelationships between different types of plant stresses.

2.3 HEALTH EFFECTS

(a) Irritant properties

Sulphur dioxide and its derivatives differ markedly from the oxides of nitrogen (but less so from ozone and photochemical oxidants) in that they produce strong irritation upon the eyes and also within the nasal passageways. Indeed, sulphur trioxide on a molar basis produces more than four times the irritant response than sulphur dioxide because it combines immediately with water to form sulphuric acid (Reaction 2.1). Sulphate in particulates can also cause similar irritation and the intensity is dependent upon concentration of the particulates and their sizes. Those particles smaller than 1 μm are the most irritating but larger particles as well as high concentrations of gaseous sulphur dioxide may also induce an involuntary coughing reflex. This draws the attention of the sufferer to the immediate hazards of the surrounding atmosphere although there is a wide variation in susceptibility to this response between individuals. Likewise sulphur dioxide can also be detected by smell by certain individuals rather better than others although at very high concentrations above 3 ppmv this sense of smell is quickly paralysed in all.

Routes of entry of sulphur dioxide and its allies into body tissues also differ from those associated with oxides of nitrogen, ozone and photochemical oxidants because more than 90 % of inhaled sulphur dioxide is absorbed in the airways above the larynx (voice-box). During severe short-term exposures, the sulphate and sulphite anions formed on the moist cell surfaces of the nasal linings penetrate the mucosal cells and bind to granules within mast cells. This causes the local release of histamine which acts as a local modulator to cause constriction of the airways and initiate local inflammation responses. At the same time, ciliary cells release mucus which serves to carry away some of the harmful anions as the nose is blown or one coughs. Tears from the eyes similarly provide some relief.

Although sulphur dioxide is the oldest recognised harmful air pollutant to man, in recent years other priorities (e.g. photochemical oxidant damage) have tended to play down the consequences to human health of this acidic gas. Because human physiological responses to sulphur dioxide such as those described above are absent from experimental animals, it has always been difficult to extrapolate results obtained from animals to the human condition but the effects of sulphur dioxide on humans are even less certain due to the protective nature of these responses. This lack of correlation, however, has not stopped a large number of animal studies over a long period (nearly 50 years) which have been mainly used to elucidate mechanisms of action and to evaluate total risk even if the risk to human health cannot be quantified accurately.

The US Threshold Limit Value (or OEL in the UK) for five 8-hour periods of a working week set at 5 ppmv appears to offer very little if any margin of safety in man. This inconsistency is due to the fact that the studies which were largely used to establish this limit showed that concentrations of fifty times ambient levels were said to produce 'little' distress in experimental animals. Consideration of Table 2.3 will show that levels of sulphur dioxide well below 5 ppmv have been documented as causing adverse human responses. It would appear that only healthiest non-smoking adults would be covered by this US Threshold Limit Value and no margin of safety is allowed for interactions caused by other air pollutants, aerosols or particulates.

(b) Industrial and urban hazards

Workers in certain industries (e.g. smelting, sulphuric acid production, etc.) who have been exposed to rather more sulphur dioxide than other groups are often claimed to have acquired a degree of tolerance to this gas. On the other hand this group have also accounted for most of the industrial fatalities and acute injuries.

Table 2.3 Human response to different levels of sulphur dioxide

Concentration† (ppmv)	Period	Effect
0.03–0.5	continuous	Condition of bronchitic patients worsened
0.3–1	20 sec	Brain activity changed
0.5–1.4	1 min	Odour perceived
0.3–1.5	15 min	Increased eye sensitivity
1–5	30 min	Increased lung airway resistance, sense of smell lost
1.6–5	less than 6 hours	Constriction of nasal and lung passage
5–20	more than 6 hours	Lung damage reversible if exposure ceases
20 upwards	more than 6 hours	Water logging of lung passageways and tissues, eventually leading to paralysis and/or death

† Generally lower if aerosols, particulates or other pollutants are also present.

Sudden deaths resemble those due to asphyxiation but, as is more normal when death is delayed, subsequent post-mortem examination reveals oedema (enhanced water uptake into tissues), destruction of the ciliated epithelial covering of the air passageways and, more significantly, invasion of the lung by bacteria.

During accidental non-fatal exposures, victims experience inflammation of the eyes, nausea, vomiting, abdominal pain, sore throat, bronchitis and often pneumonia. Normally the lungs are (and have to be) highly sterile but this predisposition to failure of the natural antibacterial mechanisms points the way to emergency treatment of sulphur dioxide-exposure cases. Prompt use of oxygen and bronchodilators along with heavy doses of antibiotics prevent permanent damage and may even make the difference between life or death.

Mercifully, sudden accidental acute cases are rare although the possibilities of chronic poisoning by sulphur dioxide over much longer periods at lower levels are greater for the general population. The literature which deals with and confirms the correlation between chronic chest disease and levels of sulphur dioxide in urban air is vast but the evidence to suggest that chronic exposure to sulphur dioxide and particulate matter plays a part in the cause and development of chronic respiratory disease has only been established with

Table 2.4 Summary of the important epidemiological surveys into the relationship between long-term health and sulphur dioxide plus particulates (from Economic Commission for Europe, 1984)

Country	Annual average pollutant levels		Effects noted
	SO_2 (ppbv)	particulates ($\mu g\ m^{-3}$)	
USA	9.5	135	Higher incidences of acute respiratory diseases
USA	21	180	Increased rate of respiratory symptoms; decreased lung function
UK	38	200	Elevated incidence rates of respiratory problems
Poland	48	270	More chronic bronchitis and asthmatic disease in smokers
USSR	48	285	Enchanced symptoms of respiratory disease
UK	86	360	Higher frequency of respiratory symptoms; decreased lung function in children

difficulty. The problems reside in the multitude of other factors which often cannot be fully excluded in large-scale epidemiological studies. Adequate allowance must be made for previous medical history, tobacco smoking, age, sex, family size, racial origin, socio-economic, seasonal and meteorological variables. The classic case of deaths of Londoners during the 'smog' of 1952 is often quoted. But, very often high incidences of death (mortality studies) and peaks in air pollution are unrelated. The average level of sulphur dioxide during the London 'smog' of 1952 was actually 1.7 ppmv; well inside the US Industrial Threshold Limit Value. The main findings of more careful epidemiological morbidity studies are shown in Table 2.4 alongside the annual pollutant levels at which the main symptoms were noted. It is studies like these that caused the World Health Organisation in 1979 to recommend guidelines to member countries of the United Nations of 15–23 ppbv (40–60 $\mu g\ m^-$) as an annual mean which is thought to incorporate a safety factor of between 2 and 3.

(c)   Toxicity of SO_2 and SO_3^{2-} within tissues

It is apparent that human tissues have more than adequate amounts of sulphite oxidase activity (see Sect. 2.1d) to remove bisulphite or sulphite almost as soon as it appears. The average dietary intake of bisulphite from preserved foods is thought to be less than 0.2 mmol per day; much less than that caused by internal breakdown of sulphur-containing amino-acids (about 25 mmol per day per person). The liver alone appears to have more than 200 times the capacity needed to oxidise these amounts. It has been calculated that a man breathing 5 ppmv sulphur dioxide for 8 hours would take up the equivalent of 1.3 mmol of bisulphite and sulphite during the day. The lungs have apparently more than 100 times the capacity of sulphite oxidase to remove this but such calculations of activities are spread over a whole day and throughout the whole lung. It is far from certain that intense local accumulations in the epithelial tissues of the lungs or nasal regions can be removed immediately before adverse localised damage occurs. The major molecular mechanisms involved would be those of free-radical and nucleophilic attack similar to those described earlier (Sects. 2.1d and e).

Studies with rabbits and dogs exposed to large amounts of (10–25 ppmv) radioactive ^{35}S-labelled sulphur dioxide have shown that lung sulphite oxidase activities are also insufficient to prevent other reactions in the lungs and to prevent ^{35}S-labelled bisulphite entering the blood. Bisulphite appears to be immediately bound up as ^{35}S-sulphocysteine ($RS^{35}SO_3$: Reaction 2.24) which is then either taken into the liver or into other body tissues. Because the liver has all the necessary detoxifying enzymes, it quickly returns its radioactivity back to the blood as ^{35}S-labelled sulphate which is then quickly eliminated by the kidneys. The remainder trapped in the other tissues persists for longer in the body before returning slowly to the blood and emerging in the urine.

The question that remains is: 'How do these results from animal studies carried out at high levels of sulphur dioxide relate to humans exposed to much lower amounts?' There seems to be general agreement that any genetic consequence due to mutagenic cytidine alterations (see Sect. 2.1e) are highly unlikely but enhanced levels of sulphocysteine in nerve tissues may be cause for concern. Low levels of sulphocysteine given intravenously to rats have, for example, been shown to cause brain abnormalities.

It is probably the adverse reactions within lung tissue close to the airways that have most relevance to humans. Epidemiological studies (see Sect. 2.3b) showing respiratory problems to be the main consequence of exposure to sulphur dioxide would tend to confirm this statement. Unfortunately, very few useful studies to attempt to understand the mechanism of the cause of these problems have been undertaken. The most reliable studies have used tissue cultures of alveolar macrophages (phagocytic cells associated with removal of debris) and lymphocytes (non-phagocytic cells associated with antibody production) isolated from both animal and human lungs. Such cells, when exposed *in vivo* or *in vitro* to sulphur dioxide, sulphite or bisulphite, initially show elevated levels of a plasma-membrane-bound enzyme called ATPase, as well as having depleted levels of ATP which implies a general reduction of cell capabilities. External levels of lysozyme (a general protein-degrading enzyme which is secreted by macrophages) were also increased by sulphur dioxide. The gas also appears to act as a trigger for fresh synthesis of lysozyme to take place inside the macrophages. Moreover, enhanced proteolytic capabilities expressed within lining tissues may also have undesirable effects upon the integrity of the surrounding aveolar cells rather than upon any ongoing bacterial infection.

As well as enzymic change, the fluidity of the surrounding cell membranes appears to be altered by sulphur dioxide, bisulphite or sulphite. From the nature of these changes it has been predicted that lymphocytes might not be able properly to recognise and bind antigen and therefore not be able to produce the appropriate antibodies. Alternatively, they may be unable to recognise and respond to foreign or aberrant cells. This reduced immune surveillance may give rise to the bronchitic and respiratory problems reported by the epidemiological surveys. Similar membrane changes seem to also occur in lung macrophages. Normally, lymphocytes produce a lymphokine referred to as MIF (macrophage inhibitory factor) which controls the movement of macrophages. After exposure to sulphur dioxide the macrophages lose the ability to respond to MIF and show increased migratory tendencies thus avoiding the phagocytic task identified by the lymphocytes. It is probably events like these that give rise to the long-term respiratory problems known to occur in humans.

SELECTED BIBLIOGRAPHY

Anderson, J. W., *Sulphur in Biology*. Studies in Biology no. 101. Edward Arnold, 1978, London.

Economic Commission for Europe, *Air-borne Sulphur Pollution*. Air Pollution Study no. 1. United Nations, 1984, New York.

Muth, O. H. (ed.), *Sulphur in Nutrition*. AVI Pub. Co., 1970, Westport, Conn.

Nriagu, J. O. (ed.) *Sulphur in the Environment*. John Wiley, 1978, New York.

VDI-Kommision Reinhaltung der Luft, *Sauerstoffhaltige Schwefelverbindungen Bibliographie 1974–1978*. Dusseldorf.

VDI-Berichte Nu.314, *Sauerstoffhaltige Schwefelverbindungen*. VDI-Verlag GmbH, Dusseldorf.

Winner, W. E., Mooney, H. A. and Goldstein, R. A., *Sulphur Dioxide and Vegetation: Physiology, Ecology and Policy Issues.* Stanford University Press, 1985, Stanford, Calif.

Ziegler, I., The effect of SO_2 pollution on plant metabolism. *Residue Reviews* **56**, 1975, 79–105.

Chapter 3

Nitrogen-based air pollutants

'It is a fire, it is a coale
whose flame creeps in at every hole'

From 'The Hunting of Cupid' by George Peele (1558–1597)

3.1 FORMATION AND SOURCES

(a) The Nitrogen cycle

If written only a short while ago this chapter would have been entitled
'Oxides of Nitrogen' and have presented nitrogen dioxide, NO_2, and,
to a lesser extent, nitric oxide, NO, as the predominant nitrogen-
containing atmospheric pollutants which give rise to problems. This
concept has been encapsulated in the abbreviation often used for
these two pollutant gases – NO_x. The use of abbreviations has been
deliberately kept to a minimum in this book but the use of NO_x elsewhere has had the unfortunate effect of gathering up a number of
different problems under one heading and then translating the implications revealed by the study of one component (nitrogen dioxide)
as symptomatic of nitric oxide as well, a component often in the largest
proportion. More seriously, it is never clear if NO_x includes or
excludes nitrous oxide (N_2O, dinitrogen monoxide) which is often
more than ten times as prevalent in the atmosphere as nitrogen
dioxide. If one writes of 'oxides of nitrogen', a much better collective

expression than 'nitrogen oxides', the meaning is clear. It includes nitrogen dioxide and nitric oxides, as well as nitrous oxide and other possibilities, N_2O_3, N_2O_5, etc.; each having their own chemical properties and individual effects on living systems. In future it would be better if the use of the abbreviation NO_x was abandoned, convenient though it may be, because it induces a belief of similarity in effect within living systems which is gradually being shown not to be the case.

Not all nitrogen-based atmospheric pollutants are oxidised. Ammonia (NH3), for example, is released into the atmosphere as a result of the decay of waste products from animals or as escaping gases during the manufacture of artificial fertilisers. As its ionic product ammonium (see Sect. 3.1f) it may also be harmful to those ecosystems which are sensitive to excessive inputs of nitrogen. Releases of ammonia are covered later in this section and some of the biological consequences of ammonium deposition are dealt with in Chs. 4 and 7.

As nitrogen circulates between plants and animals and microbes in

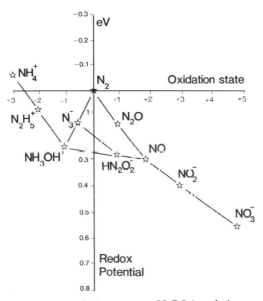

Fig. 3.1 Oxidation states of nitrogen at pH 7.0 in relation to redox potentials. Electronegativity represents a higher energetic state than electropositivity and electrons will flow naturally down such a gradient. For this reason the convention of sign when drawing such a diagram is reversed to emphasise the downwards flow. A similar concept applies to some of the subsequent figures.

the soil, wide differences exist in the reduced and oxidised components involved. In order to understand this biological nitrogen cycle, an overall appreciation of some of the complications of nitrogen chemistry is useful.

The electronic configuration about the nitrogen atom may vary widely permitting in turn a number of different oxidation-reduction states of nitrogen to exist (Fig. 3.1). These are very similar to the various forms of sulphur discussed in Ch. 2. Individual nitrogen compounds also have a number of different reduction-oxidation or redox potentials relative to the standard hydrogen electrode potential. As explained earlier (Sect. 2.1c) this is effectively a scale of relative energies whereby electrons flow from more electronegative compounds to more electro-positive compounds, releasing energy as they do so. Conversely, energy must be added in order to convert electropositive compounds to more electronegative compounds. Consequently, the biological nitrogen cycle is dominated by the underlying principles of nitrogen chemistry (Fig. 3.2).

As the following sections will show, the impact of biological activity upon the global exchanges of nitrogen is considerable (Table 3.1) if not overwhelming. Human activities cause the combustion of fossil

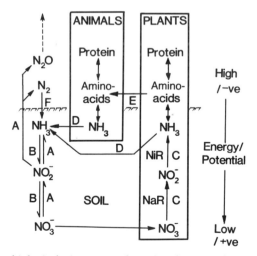

Fig. 3.2 The biological nitrogen cycle as it relates to plants, animals, the soil and atmosphere. The various processes are as follows: A, denitrification or dissimilatory nitrogen metabolism; B, microbial nitrification; C, plant assimilatory nitrogen metabolism; NaR, nitrate reductase enzyme; NiR, nitrite reductase; D, decay after death; E, digestion and F, biological nitrogen fixation. Note that the release of nitrous oxide to the atmosphere takes place mainly by denitrification.

Table 3.1 Global transfer rates of nitrogen by various processes

Process	Million tonnes of nitrogen[†] per annum
Denitrification/nitrification	160
Natural nitrogen fixation	150
Fires and combustion	70
Industrial fixation	40
Ionic exchange by rain, etc.	80
Run-off to oceans	35

[†] Which may be expressed as Mt N y^{-1} or sometimes as Tg N y^{-1} where T, tera, is 10^{12} and t, 1000 kg.

Note: Estimates of the forms and global amounts of nitrogen compounds vary but are very similar in dimension to the global rates of transfer because, with the exception of nitrous oxide which has an atmospheric lifetime of 150 years, most of the others have lifetimes measured in weeks. Because of this, the global reservoir of nitrous oxide is by far the largest (1500 Tg of nitrogen, N) followed by ammonia (1.6 Tg N), ammonium (0.5 Tg N), nitrogen dioxide (0.4 Tg N), nitric oxide (0.2 Tg N), nitric acid (0.2 Tg N) and nitrate (0.1 Tg N).

fuels or the manufacture of artificial fertilisers which then form a considerable part of this biological exchange but the bulk remains microbial.

(b) Nitrous oxide

The most abundant oxide of nitrogen in the atmosphere is nitrous oxide or dinitrogen oxide, N_2O, familiar to some in more concentrated form as 'laughing gas'. The vast majority of nitrous oxide in the atmosphere is formed by microbes in the soil by a process known as 'denitrification' (Fig. 3.3). This is an undesirable agricultural process because valuable plant nutrients are lost as certain soil microorganisms use nitrate instead of oxygen in order to respire, forming nitrous oxide and nitrogen as they do so. Soil conditions from which oxygen is excluded or to which it has poor access favour this undesirable denitrification. Consequently, stagnant water-logged or compacted soils are major sources of nitrous oxide and normally agricultural practices such as ploughing or draining discourage this anaerobic process.

Other microbes also carry out the reverse reaction known as 'nitrification' whereby conversion of ammonium or nitrite to nitrate by oxygen-dependent microbes is encouraged by aeration. Therefore

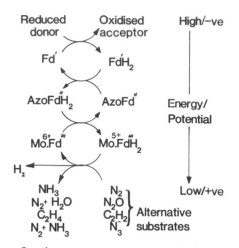

Fig. 3.3 Electron flow between components involved in biological nitrogen fixation. Abbreviations used are as follows: Fd', oxidised ferredoxin, an iron–sulphur-containing protein; Fd'H$_2$, reduced form of Fd'; AzoFd″ = an alternative ferredoxin at lower potential; Mo.Fd‴, a molybdenum-containing ferredoxin with active nitrogen fixation site. Note the alternative triple-bonded substrates for the complex as well as the release of molecular hydrogen and the change from the molybdic to molybdenous form within the MoFd complex.

one of the basic objectives of arable farming is to maximise microbial nitrification (Fig. 3.4) and at the same time, encourage the fixation of atmospheric nitrogen into ammonia by the nitrogenase enzyme complex (Fig. 3.5). By applying artificial fertilisers instead, the farmer deliberately disturbs the natural balance between these different processes and discourages nitrogen fixation. An alternative would be to plant more legumes or clovers in rotation but prevailing economics rather than good scientific practice is ultimately the deciding factor. Any excess artificial fertiliser remaining after application and not draining away is most likely to be removed by denitrification. Increasing global use of artificial fertilisers has thus resulted in an increase in the atmospheric levels of nitrous oxide because this microbial process of denitrification has been enhanced. Global production of nitrous oxide by denitrification each year amounts to about 5 % of the total atmospheric reservoir of nitrous oxide. Because nitrous oxide is a relatively unreactive molecule it has a long residence time in the atmosphere of 20 years or more. Any subsequent atmospheric increase in nitrous oxide levels due to enhanced denitrification on account of excess artificial fertilisers being applied to unsuitable land will therefore develop slowly but persist for much longer.

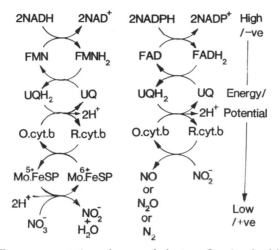

Fig. 3.4 Two representative schemes of electron flow involved in denitrification, sometimes called 'nitrate' or 'nitrite respiration'. Abbreviations are as follows: NADPH, reduced, and NADP, oxidised forms of nicotinamide adenine dinucleotide (phosphate); FMN, flavin adenine dinucleotide; UQ, ubiquinone; O, oxidised, and R, reduced forms of cytochrome *b*, cyt.*b* (a tetrapyrrole haem protein containing iron); Mo.FeSP, a molybdenum-iron-sulphur-containing protein. Generally different microbes carry out the different reactions. The flow pattern on the left which achieves a lower potential converts nitrate to nitrite and the one on the left is representative of a number of different electron flows in different organisms which produce a range of products.

Small amounts of nitrous oxide are also produced by the process of nitrification (Fig. 3.4) as a result of incomplete conversion of ammonium to nitrite. Consequently, small losses of fixed nitrogen occur at different stages during nitrification. In compensation, nitrogen fixation involves some consumption of nitrous oxide as it is an alternative substrate for the nitrogenase complex (Fig. 3.3) which is to be found in all organisms capable of trapping molecular nitrogen.

The great unknown involves the oceans of the world. The relative alkalinity of their waters as compared to those of soils on the surface of the land favour denitrification processes and, hence, nitrous oxide production because the energy yield of the denitrification reaction is inversely related to pH. This is the case especially in deep waters where oxygen is scarce or in estuaries where it may be depleted. Sewage discharges, for example, cause a general decline in oxygen content causing them to become eutrophic and therefore encourage denitrification.

Lack of removal mechanisms for nitrous oxide in the lower atmos-

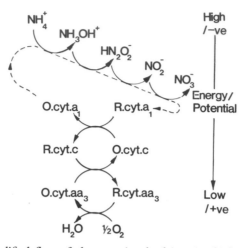

Fig. 3.5 Simplified flow of electrons involved in microbial nitrification processes. Nitrification differs from nitrogen fixation (Fig. 3.3) and denitrification (Fig. 3.4) because nitrogen compounds donate electrons instead of accepting electrons from the biological couples involved. Many different microbes are involved in nitrification processes and certain species specialise only in some of the transfers, e.g. *Nitrobacter* oxidises nitrite to nitrate. Abbreviations are as follows: Oxidised, O, or reduced, R, forms of different cytochromes (cyt.a_1, cyt.c, cyt. aa_3) having different redox potentials; cyt.aa_3, an iron-copper haem protein known as 'cytochrome oxidase'.

phere (troposphere) means that the molecules drift slowly upwards into the stratosphere where they may either undergo photolysis or reaction with atomic oxygen. Photolysis (Reaction 3.1) produces molecular nitrogen and oxygen which has no major implications but Reaction 3.2 which forms nitric oxide has a strong catalytic influence upon reducing ozone levels in the stratosphere (see Ch. 5). This link between increasing global use of nitrate fertilisers, enhanced denitrification and possible increase in the movement of nitrous oxide through the troposphere to the stratosphere has been the cause for some concern and conjecture. At the moment there are insufficient data from stratosphere studies on which to base truly reliable predictions or with which to formulate future policies (see, for example, Sect. 6.2a). It is a potential problem which has a long time period which will be difficult if not impossible to reverse if changes in the ozone cover of the stratosphere due to this cause are ever detected. The discouraging of the use of nitrate-based fertilisers is bound to be unpopular, especially with developing countries. Although still a pipe-dream of genetic engineering and biotechnology, transfer by genetic manipulation of mechanisms of nitrogen fixation, availability or uptake from non-economic to economic plants such as cereals and grasses may yet become a reality to retrieve the situation.

$$2N_2O \xrightarrow{\text{light}} 2N_2 + O_2 \qquad (3.1)$$

$$N_2O + O \longrightarrow 2NO \qquad (3.2)$$

(c) Combustion

Nitric oxide is formed by the high temperature combination of atmospheric nitrogen and oxygen, with a lesser contribution coming from nitrogen-containing components in the fuel. Reaction 3.3 is often quoted as the major process but it actually has two components (Reactions 3.4 and 3.5) involving highly reactive atoms created by the high temperature of the flame, the second much faster than the first. In the fuel-rich region of a flame, highly reactive hydroxyl ($OH^•$) radicals (see Appendix, Sect. 8.2) form nitric oxide (Reaction 3.6) but hydrogen cyanide, ammonia or amine derivatives have also been identified as precursors of nitric oxide. Odd as it may first appear, the amount of fuel nitrogen converted to nitric oxide decreases as nitrogen content of the fuel rises and the removal of sulphur from fuels increases the amount of nitric oxide produced. Therefore the burning of heavy oil, which usually has a low nitrogen content, could lead to nitric oxide emissions as great as those from nitrogen-rich coal-fired power plants.

$$N_2 + O_2 \longrightarrow 2NO \qquad (3.3)$$

$$N_2 + O \longrightarrow NO + N \qquad (3.4)$$

$$N + O_2 \longrightarrow NO + O \qquad (3.5)$$

$$N + OH^• \longrightarrow NO + H \qquad (3.6)$$

Much can be done to reduce nitric oxide emissions from furnaces in terms of design of fuel, burners and operating conditions. For example, the particle or droplet sizes of the fuel may be reduced which has the effect of providing larger surface areas to encourage more complete combustion (Reactions 3.7 to 3.9). Other parameters which may be adjusted so as to minimise nitric oxide formation are the temperatures or pressures, the availability of oxygen, the mixing

$$C + 2NO \longrightarrow CO_2 + N_2 \qquad (3.7)$$

$$2C + 2NO \longrightarrow 2CO + N_2 \qquad (3.8)$$

$$2CO + 2NO \longrightarrow 2CO_2 + N_2 \qquad (3.9)$$

rates and the residence times. Indeed it may be much cheaper (from six to sevenfold) to redesign burners to reduce emissions of oxides of nitrogen (and hence acidity) than to desulphurise retrospectively the emissions in order to reduce total acid-forming emissions.

Another approach exploited very effectively in Japanese power plants has been to inject ammonia into the combustion gas by a process known as 'selective catalytic reduction'. Reaction 3.10 is only efficient at temperatures below $1000°C$; above this the ammonia tends to form nitric oxide once again.

$$4NO + 4NH_3 + O_2 \longrightarrow 4N_2 + 6H_2O \qquad (3.10)$$

In most developed countries, road transport contributes about 30 % of the total emissions of oxides of nitrogen as opposed to 45 % from power plants and 25 % from domestic and general industrial sources. Improvements to fuel combustion by redesign of car engines and exhaust systems in order to reduce emissions of nitric oxide are possible and legislation will be introduced progressively in more countries. Alterations to the nature of the combustion within the car engine itself are most likely to go hand-in-hand with improved catalytic conversion of exhaust gases. Changes in the ratio of air-to-fuel used by engines appear to offer the best way forward in the future. Justification for the development of this 'lean-burn' concept is well shown by the gaseous changes in Fig. 3.6. Other techniques such as 'spark retardation' and 'exhaust gas recirculation' are shorter-term alternatives which have the disadvantage that they could be more easily circumvented by owners. Expensive three-way platinum/rhodium catalysts can be incorporated into car exhaust systems to encourage all three removal reactions (Reactions 3.11 to 3.13) rather than the cheaper oxidation catalysts now required by some countries. The latter only carry out Reactions 3.11 and 3.12 but have little effect on nitric oxide emissions. The problem with three-way catalysts is that removal of unburnt hydrocarbons and carbon monoxide requires plenty of oxidant but removal of nitric oxide is hindered by excess oxidant (see Fig. 3.7). Only a narrow window of oxidant availability exists for all three conversions to occur and this is difficult to maintain throughout different driving conditions and the lifetime of the catalytic exhaust system. All systems involving lean-burn car engines, three-way catalytic exhausts and exhaust recirculation also require much greater control over the fuel input to the engine over a wide

$$2CO + O_2 \longrightarrow 2CO_2 \qquad (3.11)$$

$$(H_2C) + O_2(+2NO) \longrightarrow CO_2 + 2H_2O(+N_2) \qquad (3.12)$$

$$2CO + 2NO \longrightarrow 2CO_2 + N_2 \qquad (3.13)$$

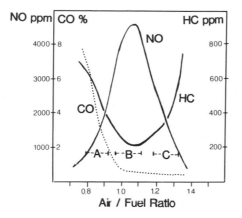

Fig. 3.6 Emissions of carbon monoxide, CO, nitric oxide, NO, and unburnt hydrocarbons, HC, from internal combustion (Otto) engines operated at different air/fuel ratios. A, range of mixtures used in the past by older engines; B, present-day engines; C, so-called 'lean-burn' engine designs of the future.

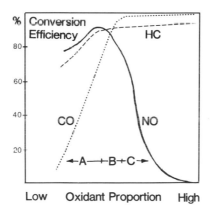

Fig. 3.7 Percentage conversion efficiencies of carbon monoxide, CO, nitric oxide, NO, and unburnt hydrocarbons, HC, by catalysts to be included in car-engine exhaust systems. The effective regions of either reduction catalysts, A, or oxidation catalyst, C, of single- or dual-bed systems presently in use are to either side of region B which is the effective range of platinum/rhodium three-way catalytic systems still under development.

range of driving conditions. Computerised fuel injection with feed-back loops from exhaust sensors will therefore become standard and inevitably increase costs.

Legislation favouring exhaust improvements also has a beneficial effect in that fortuitously it has or will hasten a general reduction in lead content of fuels and hence lead emissions to the atmosphere. This is necessary because lead would quickly poison the exhaust catalysts themselves.

Other industrial processes produce significant emissions of oxides of nitrogen. Ammonium nitrate is a widely used nitrogenous fertiliser for agriculture and is produced world-wide by a large number of factories. It is usually synthesised from the reaction of ammonia with nitric acid (Reaction 3.14). Ammonia is generated by the Haber process (Reaction 3.15) whilst nitric acid is formed from the catalytic oxidation of ammonia (Reactions 3.16 to 3.18). Escapes and emissions of oxides of nitrogen or ammonia, as well as particulate ammonium nitrate, to the atmosphere from such factories may be significant and often give rise to considerable local pollution.

$$HNO_3 + NH_3 \longrightarrow NH_4NO_3 \tag{3.14}$$

$$N_2 + 3H_2 \longrightarrow 2NH_3 \tag{3.15}$$

$$4NH_3 + 5O_2 \longrightarrow 4NO + 6H_2O \tag{3.16}$$

$$2NO + O_2 \longrightarrow 2NO_2 \tag{3.17}$$

$$3NO_2 + H_2O \longrightarrow 2HONO_2 + NO \tag{3.18}$$

(d)   Atmospheric oxidations

Once in the atmosphere, nitric oxide is rapidly oxidised to nitrogen dioxide mainly by reaction with ozone (Reaction 3.19) which is counter-acted by a photochemical back-conversion (Reaction 3.20). As Ch. 5 shows, ozone may be regenerated by Reaction 3.21 in the presence of a surface, M, which serves to carry away excess energy and assist in the formation of new bonds. As a consequence of this catalytic cycle (i.e. Reaction 3.19 working against Reactions 3.20 and 3.21), a photostationary state is established which permits significant concentrations of nitric oxide, nitrogen dioxide and ozone to exist in the presence of each other; the proportions of which are to some extent controlled by light, temperature and the presence of particulates and hydrocarbons (see Sect. 5.1).

$$O_3 + NO \longrightarrow NO_2 + O_2 \tag{3.19}$$

$$NO_2 + light \longrightarrow NO + O \tag{3.20}$$

$$O + O_2 + M \longrightarrow O_3 + M \tag{3.21}$$

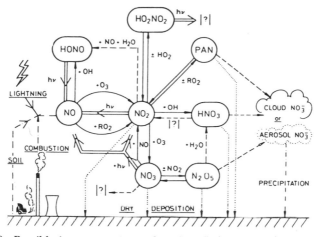

Fig. 3.8 Possible interconversions of oxides of nitrogen in the atmosphere. (Courtesy of Drs. Cox and Penkett, AERE, Harwell, UK and D. Reidel. Publ. Co., Dordrecht, The Netherlands.)

Apart from photolytic breakdown (Reaction 3.20), a variety of alternative reactions may be responsible for the removal of nitrogen dioxide ((Fig. 3.8). In bright light, photochemically generated hydroxyl radicals or ozone may produce nitric acid (usually represented as HNO_3 but $HONO_2$ is structurally more correct) by means of Reactions 3.22 or Reactions 3.23 to 3.25 which return hydroxyl radicals back into the system (Reaction 3.26). Other products stemming from the formation of highly reactive nitrogen trioxide include nitrogen dioxide and its equilibrium alternative, nitrogen tetroxide (Reactions 3.27 and 3.28) or more nitric oxide by the action of light (Reaction 3.29). Reactions 3.23, 3.27, 3.28 and 3.29 therefore assist Reactions 3.19 and 3.20 in maintaining the major equilibria between NO and NO_2 in the atmosphere. Current estimates predict that each nitric oxide, radical destroys on average 300 ozone molecules (Reaction 3.19) before it is trapped.

$$OH^\bullet + NO_2 + M \longrightarrow HONO_2 + M \qquad (3.22)$$

$$NO_2 + O_3 \longrightarrow NO_3 + O_2 \qquad (3.23)$$

$$NO_3 + NO_2 + M \longrightarrow N_2O_5 + M \qquad (3.24)$$

$$N_2O_5 + H_2O \rightleftharpoons 2HNO_3 \qquad (3.25)$$

$$HONO_2 + light \longrightarrow OH^\bullet + NO_2 \qquad (3.26)$$

$$NO_3 + NO \longrightarrow 2NO_2 \rightleftharpoons N_2O_4 \qquad (3.27)$$

$$NO_3 + NO \longrightarrow NO_2 + O \qquad (3.28)$$

$$NO_3 + light \longrightarrow NO + O_2 \qquad (3.29)$$

Other important atmospheric reactions involve the formation of both peroxyacyl nitrates (PANs: Reaction 3.30) and peroxynitric acid (HO_2NO_2). Most PANs have a half-life of about 1 hour but peroxynitric acid is more unstable and lasts only a few seconds in the lower troposphere. However, they are all thermally stable in the colder upper regions of the troposphere and lower stratosphere where interactions with hydroxyl radicals are more prevalent. Formation of peroxynitric acid relies on the presence of the hydroperoxyl radical ($HO_2^{\bullet}$) which is formed by the rapid addition of atomic hydrogen to molecular oxygen. This radical and the conditions for its formation (i.e. ionising radiation) are again more likely in the stratosphere.

$$CH_3CO.O_2 + NO_2 + M \longrightarrow CH_3COO_2NO_2 + M \quad (3.30)$$

Nitrous acid (HONO rather than HNO_2) may also be formed very rapidly in the gas phase by reaction between nitric oxide, nitrogen dioxide and water (Reaction 3.31). Furthermore, nitrous acid will also decompose in aqueous solution by a series of interconversions (Reactions 3.32 to 3.34). However, this decomposition to nitric oxide and nitric acid is slowed down in dilute solutions because Reaction 3.32 forming nitrogen trioxide only occurs in concentrated solutions.

$$NO + NO_2 + H_2O \longrightarrow 2HONO \qquad (3.31)$$

$$4HONO \longrightarrow 2N_2O_3 + 2H_2O \qquad (3.32)$$

$$2N_2O_3 \longrightarrow N_2O_4 + 2NO \qquad (3.33)$$

$$N_2O_4 + H_2O \longrightarrow HONO + HONO_2 \qquad (3.34)$$

The presence and composition of oxides of nitrogen in the atmosphere is therefore very complicated. A multitude of different reactions (see Fig. 3.8) are influenced to a greater or lesser extent by temperature, light and different levels of dilution because all of these molecules, atoms or radicals have to collide in order to react. The chances of this are reduced as pressure falls with increasing height above the Earth but, as molecules rise, some of the intervening collisions with unreactive molecules are reduced and radiation levels increase. Consequently half-lives of reacting species vary at different heights and some intermediates which exist for very short periods at ground level persist for long periods in the stratosphere. On the other hand, those species which are relatively inert at ground level have a chance to be slowly transported upwards into the stratosphere. Nitrous oxide is a good example of this possibility.

(e) Dry deposition

All oxides of nitrogen may be removed from the atmosphere by absorption at ground level by earth or into the sea or they may be taken up by vegetation. These direct processes are known collectively as 'dry deposition'. The velocity of deposition of each atmospheric gas differs widely and depends upon the nature of the uptake surfaces. Rates for chemical species like nitric acid are very high because they are so reactive whilst the deposition velocity of nitrogen dioxide is lower than that of sulphur dioxide or ozone (Table 3.2) but much higher than that for nitric oxide (Table 3.2) or nitrous oxide. Estimates of global wet- and dry- deposition rates of both nitrate and nitric acid (which are difficult to separate) amount to about 25 million tonnes of nitrogen per annum whilst dry deposition of nitrogen dioxide accounts for an additional third of this amount.

Measurements of deposition rates are complicated because the gases may interact in many different ways. For example, undissolved nitric acid may also form ammonium nitrate aerosol (Reaction 3.14); a salt which has quite a high dissociation constant so that significant equilibrium levels of both ammonia and nitric acid are present within such aerosols. Unfortunately, very little information exists concerning the deposition velocities of nitric acid or aerosol nitrate. As may be expected, dry-deposition rates of particles range over several orders of magnitude depending upon the sizes of the particles (see Sec. 1.7) as well as the degree of turbulence within the atmosphere.

Table 3.2 Deposition velocities (mm s^{-1}) of a number of atmospheric pollutants towards different surfaces

Surfaces	Pollutant				
	NO	NO$_2$	PAN	SO$_2$	O$_3$
Soils	1.9	1.6	2–30	2–11	2.5–10
Seawater	0.015	0.15	0.2	2	0.5
Freshwater	0.007	0.1	—	1	0.1
Plants	1	4–60	6	1–29	1–17

(f) Ammonia volatilisation

Ammonia, NH$_3$, a highly pungent gas, is a raw material used by industry for the synthesis of ammonium nitrate fertiliser (see Sect. 3.1c) as well as for the synthesis of plastics, explosives, dyes and drugs. It also used to have considerable use as a refrigerant gas but is still produced in considerable quantities during the incineration of waste material (particularly plastics) and around oil refineries. All these local industrial sources of ammonia pale into insignificance on a global scale when biological decay processes are considered (see later). Atmospheric concentrations in temperate rural regions range from 5 to 10 ppbv but are much higher near the Equator (around 200 ppbv). In urban regions, levels may rise to 280 ppbv but much higher concentrations may be recorded (up to 10 ppmv) close to industrial sources. Concern has been growing that global atmospheric levels of ammonia may be rising quite quickly owing to the accelerated use of artificial fertilisers in agriculture.

The ubiquitous process of ammonia release from the biological degradation of proteins in soil organic matter, plant residues and animal wastes into the atmosphere is known as 'ammonia volatilisation'. There are a variety of factors which affect the rate of this process and some of them are far from simple and not merely a consequence of being at the end of a metabolic decay process which produces ammonia. This is because ammonia as a basic gas very readily forms ions, complexes or compounds of varying stability. Foremost amongst these are the affinity of ammonia for water to form ammonium ions, a process that is strongly affected by the pH of the solution (Reaction 3.35). As Ch. 4, emphasises there is much more likelihood of ammonia volatilisation at high pH (or alkaline) values. Other chemical species, such as carbon dioxide, or changes in temperature may also have a marked influence upon this ammonia-ammonium relationship because both the water ionisation and ammonia dissociation constants are temperature-dependent. Furthermore, the partitioning of ammonia between aqueous and gaseous phases also varies markedly with temperature.

$$NH_3 + H_2O \rightleftharpoons NH_4^+ + OH \hspace{3em} (3.35)$$

Reactions of ammonia with substances other than water also occur. In soils, ammonia may be adsorbed on to both clay and organic particles or it may react with carbonyl and other acidic groups to form exchangeable salts. Alternatively, it combines with other organic (particularly phenolic) components to form non-exchangeable products. This means that different soils have different rates of ammonia volatilisation and these, in turn, are affected by their relative water contents. Movement of ammonia in the gaseous phase through soils is via tortuous diffusion pathways but in the aqueous phase it may be transported either by diffusion or by convection if the water is moving relative to the soil. Generally, the diffusion coefficient increases with both increasing soilwater content and ammonium concentration but flooding produces complications as there is more chance that the soil beneath is anaerobic and therefore more likely to carry out denitrification (see Sect. 3.1b).

Formation of ammonia from ammonium ions (Reaction 3.35 in reverse) also removes hydroxyl ions. Therefore, as ammonia is lost to solution, the medium becomes acidified at rates dependent on the buffer capacity and, consequently, volatilisation of ammonia may be more likely from soils where the acidity produced can be neutralised by high levels of carbonate or other forms of alkalinity as Sect. 4.2(a) later shows. To a certain extent this explains why larger emissions of ammonia are caused on calcareous soils or after liming when ammonium-based fertilisers are also applied.

Various other processes are implicated in volatilisation because ammonia loss varies with the anion present in the fertiliser. Volatilisation increases as the solubility of the non-nitrogenous reaction product decreases. For example, ammonium fluoride, ammonium sulphate or diammonium phosphate react with calcium carbonate to form calcium fluoride, sulphate or phosphate. All of these products have relatively low solubilities and it is precipitation of calcium sulphate that drives Reaction 3.36 in favour of ammonium carbonate which then hydrolyses (Reaction 3.37). However, if an equilibrium is allowed to establish itself by means of a soluble calcium salt (e.g. calcium carbonate, nitrate or chloride) then a much lower concentration of ammonium carbonate is formed (Reaction 3.38) which then reduces the amount available for decomposition (Reaction 3.37).

$$(NH_4)_2SO_4 + CaCO_3 \longrightarrow CaSO_4(ppt) + (NH_4)_2CO_3 \qquad (3.36)$$

$$(NH_4)_2CO_3 + H_2O \longrightarrow 2NH_3 + CO_2 + 2H_2O \qquad (3.37)$$

$$2NH_4NO_3 + CaCO_3 \rightleftharpoons Ca(NO_3)_2 + (NH_4)_2CO_3 \qquad (3.38)$$

As an added complication, urea is a frequently used as an alternative artificial fertiliser because the enzyme urease, which is widely distributed in plants, microbes and soils, catalyses hydrolysis to bicarbonate and ammonium (Reaction 3.39). As urease activity tends to be greater in soils with heavy organic contents and rather less in calcareous soils, the usual relative rates of ammonia volatilisation are reversed. Similar enhancement of ammonia release from land treated with animal slurry or wastes also occurs as the ammonium bicarbonate decays (Reaction 3.40). However, the observed ratio of ammonia to carbon dioxide release (1 : 1) by Reaction 3.40 from slurry-treated

$$\begin{matrix} NH_2 \\ \diagdown \\ \quad\ C = O + H_2O \longrightarrow HCO_3^- + 2NH_4^+ \quad (3.39) \\ \diagup \\ NH_2 \end{matrix}$$

$$NH_4HCO_3 \longrightarrow NH_3 + CO_2 + H_2O \qquad (3.40)$$

fields is half that of Reaction 3.37 on ammonium-fertilised land although similar temperature and pH dependencies hold for both processes involved in the release of ammonia. Nevertheless, the cation-exchange capacities of such slurry-treated soils is much less of a controlling factor largely because most of the ammonia volatilisation occurs directly from the slurry surfaces and not from the soil beneath. Large losses of ammonia also arise from stored animal manures in stockyards or from human sewage works. The factors involved are similar to those of applied slurries. Thus periodic addition of fresh material to the tops of piles of manure or the agitation of sewage ponds greatly accelerate the rates of ammonia volatilisation.

Actual levels of ammonia in the atmosphere have been shown to be mainly due to the direct hydrolysis of the urea in animal urine; other contributions being secondary to this major cause. In Europe as much as 10 % of potential nitrogen is lost directly by ammonia volatilisation and in warmer climates this can rise to as much as 30 %. The amount of nitrogen released into the atmosphere globally by volatilisation is huge – between 115 and 245 million tonnes of nitrogen per annum. Any increases that have occurred over the past two decades must therefore be due to: (i) increased animal stocking levels; (ii) increased human population; (iii) increased use of artificial fertilisers either in the form of ammonium nitrate or urea; or (iv) decreased sinks for ammonia and/or ammonium uptake. The first three go hand-in-hand. As material standards improve, humans move from plant-orientated to animal-based diets. This means more food has to be grown to feed animals, which can only be done by using more artificial fertiliser. The influence upon sinks is a separate subject in its own right and is considered in Ch. 7. Nevertheless, much could be done to reduce losses associated with applications of nitrogen fertilisers. Direct injection of anhydrous ammonia or urea at the right depth into the soil has not been extensively exploited. Even substituting urea for ammonium in irrigation waters (which do not have significant urease activities) will reduce losses due to ammonia volatilisation below 2% in poorer regions of the world where nitrogen is unduly expensive and losses by this route tend to be higher in tropical countries because of the heat.

3.2   PLANTS

(a)   Access

Movements of water vapour from a plant and of carbon dioxide into a leaf have been examined in great detail by plant scientists. Research so far has indicated that most pollutant gases like sulphur dioxide and ozone move into a leaf in a manner similar to that of carbon dioxide (see Sect. 2.2a). It was thought for a time that the movement of the oxides of nitrogen into leaves was subject to similar diffusive resistances as had been found for sulphur dioxide but recent research has thrown doubt upon this being the only pathway for oxides of nitrogen. Recent calculations have shown that cuticular resistances for nitrogen dioxide are very much lower than those for sulphur dioxide and ozone. Thus, even when stomata are closed, significant amounts of oxides of nitrogen may enter a leaf through the epidermal layers. If this is proved conclusively to be the case then plants with closed stomata in bright sunlight responding for example to water deprivation could be especially vunerable to damage by oxides of nitrogen.

Considerable air spaces and exposed cell wall surfaces exist within a leaf. As a consequence, the inner cells of a leaf mesophyll (especially towards the underside, i.e. the spongy mesophyll) are generally 50–80 % air by volume and the exposed surface area of the cell walls within this mesophyll layer is considerably larger than the external surface area of a leaf. Cell walls surrounded by extracellular water therefore provide a very large surface area for the absorption of oxides of nitrogen. Once inside the leaf, the solubility of the oxides of nitrogen in the extracellular fluid is therefore an important factor in determining the rate of uptake. Both gaseous nitrogen dioxide and, especially, nitric oxide are only slightly soluble in water but nitrogen dioxide increases its apparent solubility vastly by reacting with water to produce nitric acid (Reaction 3.18).

The problem is complicated by the fact that extracellular fluid is not the same as simple water. Little is known about this fluid as it occurs in such small quantities and hence even more difficult to study in isolation. It is known that it has a complex composition which includes many organic and inorganic ions as well as some quite large molecules and is buffered to a certain extent around or just below neutrality (pH 5–7). This means that information from simple uptake studies of nitrogen dioxide into unbuffered water may not reflect what actually happens on the surface of a cell inside a leaf. Some studies have shown that gaseous nitric oxide undergoes a comparatively slow reaction with water maintained at pH 7 to form dissolved or solvated nitric oxide molecules which then react with themselves to give both nitric and nitrous acids. Far more likely is the persistence of nitrous acid (HONO rather than HNO_2) in dilute solutions formed by Reaction 3.31 but not fully decomposed because Reaction 3.32 is unfavourable when diluted (see Sect. 3.1d). Normally, nitrous acid is oxidised in solution to nitric acid by ozone, hydrogen peroxide and other oxidising agents but it may be converted back to nitric oxide when reducing agents like ferrous iron are present. Therefore the amount of nitrous acid persisting and available for movement into the cell from the extracellular fluid will depend to some extent on the presence of other gases and the nature of the dissolved components in this phase.

Gaseous nitric oxide is actually taken up by plants about a third as well as atmospheric nitrogen dioxide. The very low solubility of this gas in water would appear to contradict this possibility but once again the explanation of this discrepancy must reside in differences between water and extracellular fluid. There is some evidence, for example, that nitric oxide may form a hydrated species in solution or that it may react directly with metal cations such as iron or with certain organic molecules.

A common technique used in biology is to tag a particular molecule with a marker that can be used to trace the movement of that substance or its derivatives through different phases and compartments of a biological system. Very frequently this is done by choosing a radioactive isotopic form of an atom that would normally be non-radioactive and, by following the characteristic radiation, track the fate of the compound containing the tag. Unfortunately, the natural world does not have a convenient radioactive isotope of nitrogen. The one that exists for longest is ^{13}N which has a half-life of only 10 min. Studies using oxygen are similarly hampered. In such cases, non-radioactive isotopes of nitrogen and oxygen are used and the labelling followed by differences in mass of the isotopes using a mass spectrometer. This is technically more difficult, less convenient and often not so discriminating. However, in experiments using both ^{15}N-labelled nitric oxide gas and ^{18}O-labelled nitrate in solution, the ^{15}N has been shown to move into both nitrite and nitrate forms and the ^{18}O into nitric oxide. Thus in certain solutions that contain both nitric oxide and nitrate ions, nitric oxide is capable of forming both nitrate and nitrite ions and vice versa.

The extracellular fluid within a leaf also contains a variety of dissolved gases, undissociated acid and various ions. Nitric acid (a strong acid) will be completely ionised to protons and nitrate ions whilst nitrous acid (pK_a 3.3) will form both nitrite ions and protons but, at the pH of extracellular fluid (pH 5–7), a considerable proportion of undissociated nitrous acid will remain. Some of this, along with the nitrite ions, may be oxidised whilst in this phase.

Before access from this phase to the cell contents is achieved two more barriers exist – one more than in animals. The cellulose cell wall below is a very complicated matrix. A dense mass of cellulose polymer fibrils provide a structural framework against which the turgid cells press and by doing so provide a large element of the structural rigidity of plant tissues. Within the milieu of these polymers, ions or undissociated molecules, gases diffuse with differing degrees of ease towards the cell membrane beneath. Often given different names, such as 'plasma membrane' or the 'plasmalemma', the bounding cell membrane is yet another complex environment with both water-attracting (hydrophilic) and water-repelling (hydrophobic) phases. Some molecules are permitted to enter passively, some are actively taken up by specific uptake or active mechanisms, and others are not able to enter at all. In the case of the products of oxides of nitrogen in solution, this appears to be a passive entry process within plants but it has been little studied.

If, conceptually, the processes by which oxides of nitrogen enter the extracellular fluid, cross the cellulose wall, the cell membrane and the cell to the site of action, are linked as a series of resistances (collectively called the 'mesophyll resistance', see Ch. 2) then this may be added to the other resistances such as boundary layer resistance as one of the major influences upon the uptake of oxides of nitrogen into plants (Fig. 2.11). In the past, because it was relatively easy to measure boundary and stomatal resistances but very difficult if not impossible to measure mesophyll resistance accurately, it was often convenient to ignore or treat the latter as non-existent. This view has been seriously questioned as there is no evidence to suggest that it is very small by comparison to the other two. Indeed some of the mechanisms which make a plant either tolerant or sensitive to the effect of a pollutant may be ascribed to them having differing mesophyll resistances. Those that make it more difficult for the products of the oxides of nitrogen in solution to enter will be at an advantage over those that permit easy access. As we shall see later, there exist a number of mechanisms in a plant cell which induce the removal of the products of pollutants, especially those caused by oxides of nitrogen. If these removal mechanisms are efficient then, using an electrical analogy, the current flow will be greater if the resistance stays the same. However, if some control on either the metabolism or diffusion is exerted, the resistance may change and the flow to some extent will be controlled. Understanding both the variable and residual elements of the biochemical components that make up this mesophyll resistance may well be a better way of analysing just how plants cope and respond differently to the inward flux of the products of oxides of nitrogen in solution. As yet there has been little progress towards doing this but it is obviously an important objective for the future.

(b) Root uptake

Two major pathways in plants have been demonstrated with tracer experiments using ^{15}N-labelled nitrogen dioxide. These have shown that direct incorporation through the leaves (and stems), as well as uptake by roots after nitrogen dioxide has been absorbed into the soil, are competing processes. Atmospheric nitrogen dioxide absorbed by the soil is converted to nitrate and nitrite by microbial processes. Increases in ammonium concentration of NO_2-exposed soils may also occur and it seems that water-logging of the soil and compaction are important factors in determining the amounts and location of these three ions by virtue of alterations in the balances mentioned earlier between nitrification and denitrification processes within the soil (Sect. 3.1a). Nitrogen derived from nitrogen dioxide can be taken up by roots and metabolised into plant constituents although this process takes a long time. Even at high concentrations of $^{15}NO_2$ (4–8 ppm), the amount of ^{15}N-labelled nitrogen taken up by roots via the soil is small compared to the direct incorporation through the leaves. Consequently, the soil route might be important only under long-term exposure conditions. Preliminary investigations involving solution culture of plants have shown that this indirect route via the roots could involve a very substantial input of nitrogen derived from atmospheric sources of nitrogen dioxide.

The whole question rests on what alternative inputs of nitrogen exist. For soils which are intensively cultivated and have heavy additions of artificial or natural fertilisers applied to them, the extra availability of nitrogen from the atmosphere is insignificant. However, large areas of land, such as forest, scrub or tundra do not have these additions. Under these limiting conditions the extra nitrogen entering these ecosystems becomes a major input especially when nitrogen for more growth is scarce and often unavailable. Environmental conditions around roots therefore influence the response of plants to nitrogen dioxide but factors such as water availability, soil temperature and mineral nutrient supply are also important. They may alter the physiological status of the plant and therefore also influence the sensitivity of the plant to various pollutants.

The soil nitrogen supply can influence the net flux of nitrogen dioxide into plants and the effects of exposure upon growth. Under low levels of soil nitrogen, plants exposed to low concentrations of oxides of nitrogen may even show stimulated growth. For example, certain plants (e.g. grasses) grown on nitrogen-deficient soil and exposed to oxides of nitrogen show a slight improvement in yield whereas plants from the same stock but grown with an adequate supply of nitrogen fertiliser have a decreased yield and sometimes show evidence of visible damage. It seems fairly certain that higher nitrogen availability results in changes in the susceptibility of plants towards atmospheric levels of oxides of nitrogen as well as by affecting growth. Similar results have been found with other pollutants such as sulphur dioxide, ozone and ammonia in the presence or absence of nutrients like sulphur, phosphorus and potassium. On balance it would appear that in order to obtain an actual increase in growth and yield in grasses or cereals due to pollution or even to minimise the damage caused, the nutrient availability would have to be so low as to produce deficiency symptoms thus eliminating any possible benefits atmospheric sources of nitrogen could have for arable crops. In the case of timber, shrubs and natural vegetation, by contrast, the balance between benefit and loss may be much closer.

The level of nitrogen supply to the leaves from the roots can influence the response of plants to atmospheric nitrogen dioxide in a number of other ways. Nitrogen nutrition has been found strongly to influence the ratio between shoot and root growth rates as the root/shoot ratio decreases with increasing nitrogen supply to give increased 'leafiness' and poorer root development. Consequently, any increased availability of nitrogen to the roots would increase the amount of leaf area available for pollutant uptake relative to the whole plant biomass. Moreover, nitrogen supply to roots has been shown to influence the various diffusion resistances such as those of the stomata and the mesophyll as well as being able to exert effects upon the fixation of carbon dioxide. Photosynthesis (see later) depends on nitrogen assimilation to make essential molecules (e.g. amino-acids and proteins) and in most species the capacity for fixing carbon dioxide and transpiring water is directly related to the nitrogen content of the leaf. Additional oxides of nitrogen supplied to leaves may disturb these balances but alternative supplies of nitrogen may simultaneously modify pollutant uptake mechanisms.

Satisfactory root growth normally benefits plant shoot growth. Consequently, a variety of soil conditions may be expected to affect response of a plant to air pollutants. Unfortunately, little seems to have been done to study this possibility. Factors such as water supply and salinity, as well as mineral nutrient availability, are known to be important and to influence root/shoot ratios, diffusion pathways, photosynthesis, etc. Water stress, for example, can affect stomatal and mesophyll resistances and thereby pollutant uptake. Furthermore, the temperature of the roots is an important factor influencing both growth and carbon dioxide exchange within plants. This may be why there are differences in the responses of plants to atmospheric pollution during the summer and the winter.

In summary, much needs to be done to evaluate the direct effects of oxides of nitrogen on root uptake mechanisms. Preliminary results obtained using ^{15}N-labelled nitrate in culture solutions have indicated only a slight effect on root nitrogen uptake, whilst root activities relating to absorption (e.g. respiration) were almost the same as those in root systems of unpolluted plants. The major problems still remain. How much nitrogen from oxides of nitrogen can enter a leaf directly or come in through the soil-to-root pathway? Only by using isotopes of nitrogen (e.g. either, $^{15}NO_2$ or ^{15}NO in the presence of ^{14}N-labelled nitrate or, $^{14}NO_2$ or ^{14}NO in the presence of ^{15}N-labelled nitrate) can distinct sources of nitrogen be distinguished to show clearly how much NO_2-derived nitrogen is taken up via the leaves and how much NO_2-derived or normal nitrogen is taken up by the roots.

(c) Damage or benefit?

One difficulty experienced when discussing the toxic effects of high concentrations of oxides of nitrogen is to decide what represents a 'high' concentration. Each environment has its own 'high' concentration but in this section devoted to the majority of plants outdoors a concentration greater than 1.0 ppmv is regarded as high. Such levels of pollutant have been frequently measured in commercial glass-houses (see Sect. 3.2d) but they may also occur for short periods outdoors in urban environments.

The toxic effects of the oxides of nitrogen are directly or indirectly linked to the chemical forms of the pollutant entering the plant; the nature of which has already been discussed. Reduction of nitrite and the incorporation of ammonium into glutamate (Fig. 3.9 and 3.10, see also Fig. 2.15) requires energy supplied directly from the photosynthetic reactions. Any increase in the flux of nitrogen through the pathway inevitably withdraws both energy and carbon-skeletons away from other reactions. However, it appears unlikely that simple competition can totally explain the effects of oxides of nitrogen on photosynthesis and growth reported over the last 15 years. Indeed, increasing the carbon dioxide concentration in the air supplied to tomato plants, which are also polluted with oxides of nitrogen, produces bigger plants. If the reductant supply is limited by the oxides of nitrogen then the fixation of the extra carbon would not occur and there would have been little effect on growth. Clearly this is not the case.

Recent studies have shown that even though cellular adaptations to normal metabolism do occur, levels of nitrite and nitrate are raised at the site of likely action in the chloroplasts as a consequence of exposure to atmospheric levels of nitrogen dioxide (0.3 ppmv). Toxic effects to plants therefore may be caused by excess nitrate or nitrite, as well as by nitrous acid, ammonia and ammonium ions.

Most plant species appear able to tolerate large accumulations of nitrate even though such high concentrations may be undesirable in crop plants used directly as food by humans. On the other hand, accumulations of nitrite can have serious toxic effects on plants as well as humans. A number of investigations have concluded that it

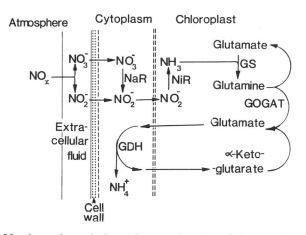

Fig. 3.9 Uptake and metabolic pathways of oxides of nitrogen into plant tissues from the atmosphere, through the extracellular fluid, cell wall, cell membrane and cytoplasm into the chloroplast. Enzymes involved include nitrate reductase, NaR, nitrite reductase, NiR, glutamine synthetase, GS, glutamate synthase, GOGAT, and glutamate dehydrogenase, GDH. For a clearer biochemical relationship between the last three enzymes see Fig. 3.10.

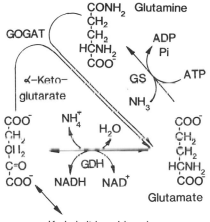

Krebs' citric acid cycle

Fig. 3.10 Interconversions between glutamate, glutamine and α-ketoglutarate (2-oxoglutarate) involving the uptake and release of ammonia in plants. The cycle involving the enzymes glutamine synthetase, GS, and glutamate synthase, GOGAT, is much more favourable, whilst cytoplasmic glutamate dehydrogenase, GDH, is much more likely to catalyse in direction of deamination.

is the acidification reactions which accompany nitrite accumulation (Reaction 3.41) which account for the toxic effects in plants but undissociated nitrous acid itself may also be harmful.

$$HONO \; \rightleftharpoons \; H^+ + NO_2^- \tag{3.41}$$

A pK_a of 3.3 (see Table 8.3, Appendix) for the equilibrium governing nitrous acid (Reaction 3.41) ensures that some undissociated acid exists in the presence of nitrite even at a neutral pH. Any accumulation of nitrite therefore will also increase the concentration of nitrous acid. Both of these chemical species can enter the chloroplast and alter the proton gradients across membranes such as those engaged in photosynthesis. It will be recalled from Ch. 2 (and Appendix, Sects. 8.3 and 8.4) that one of the most vital functions linking electron flow within photosynthetic membranes (thylakoids) to systems forming chemical energy in the form of ATP is a chemiosmotic mechanism. This depends upon the creation of proton gradients across the photosynthetic membranes so that the stroma of the chloroplast become more alkaline whilst the inside (the thylakoid lumen) becomes more acidic. Any counteracting process which causes increased acidification of the stroma will thus damage the ability of chloroplasts to form ATP effectively. The consequences could be expressed in a number of ways because ATP is required in so many different biosynthetic reactions (Figure 2.14). The most important of these, by far, is protein synthesis – an energy-demanding process which is very similar in plants, animal or microbes which often consumes as much as 80 % of all the ATP produced in such organisms.

Furthermore, any alteration in stromal pH may also affect the reactions of carbon dioxide fixation. Adding nitrite to a suspension of chloroplasts is known to cause changes in stromal pH and thereby inhibit the fixation of carbon dioxide. It has been calculated that a shift in pH from 7.8 to 7.5 inside chloroplasts would inhibit the fixation of carbon dioxide by about 40%. Experimental additions of nitrite amounting to 0.2 μM final concentration are quite capable of causing depressions of 0.3 pH units so, if such changes occurred *in vivo*, detrimental reductions in the rates of fixation of carbon dioxide are likely. These may explain some of the observed depressions of plant growth due to atmospheric levels of oxides of nitrogen.

Two more important bioenergetic processes in plants, respiration and photorespiration are also noticeably inhibited by exposure of beans to high levels of nitrogen dioxide (3 ppmv). These reductions are actually greater than any inhibition of photosynthesis under similar conditions. It is not known exactly how these processes are affected by high concentrations of nitrogen dioxide but a chemiosmotic explanation involving nitrite and nitrous acid is again the most likely. When similar experiments were done with pine at much lower and more realistic atmospheric levels of nitrogen dioxide, no inhibition of respiration was detected. Furthermore, in preparations from isolated chloroplasts or mitochondria, no changes in electron flow occurred between carriers as a consequence of nitrate or nitrite even at concentrations tenfold higher than would be expected to arise by exposure to reasonably high (0.3 ppmv) atmospheric levels of oxides of nitrogen. In retrospect, it would appear that bioenergetic explanations of growth depressions due to nitrate or nitrite alone are unlikely and that, apart from acidification (see above), other agents are more likely causes of any inhibitions.

Nitrous acid has been suggested as being capable of reacting to form nitrogen- or carbon-containing nitroso-derivatives (Reaction 3.42) but, when diluted, the slowness of Reaction 3.32 and the decreased likelihood of the NO^+ species being produced at neutral pH values would make this less likely. However, attack by nitrous acid on primary aliphatic amines such as amino-acids is a possibility (Reaction 3.43). How important this process is in a biological context has not been ascertained.

$$\begin{array}{c} R \\ \diagdown \\ NH + HONO \longrightarrow \\ \diagup \\ R \end{array} \qquad \begin{array}{c} R \\ \diagdown \\ N-N=O + H_2O + N_2 \\ \diagup \\ R' \end{array} \qquad (3.42)$$

$$RNH_2 + HONO \longrightarrow ROH + H_2O + N_2 \qquad (3.43)$$

Nitrogen dioxide is also a reactive molecule in gaseous form and will readily attack some organic molecules like lipids. These interactions are fairly complex but isomerisation of carbon–carbon double bonds from *cis*- to *trans*-orientations has been detected as well as some peroxidation (see Ch. 5). However, where lipid layers in plants have been covered with protective aqueous layers, long-term exposure to high atmospheric concentrations of nitrogen dioxide has failed to show any evidence of these reactions.

Nitric oxide is less reactive than nitrogen dioxide but could react with transition metal atoms such as ferrous iron. This possibility is greater if these already have some of their valencies satisfied by coordination linkages to sulphur atoms of ferredoxins or to aromatic ring structures such as are to be found in the pyrroles of haems. The latter, and similar structures, are not likely in extracellular fluids but are found in most cellular locations especially in mitochondria and chloroplasts. If nitric oxide does penetrate as far, or is released by other reactions (e.g. Reaction 3.35) then this type of reaction could cause some disturbance. No studies have yet been undertaken to investigate this possibility.

Reports of growth reductions caused by low concentrations of oxides of nitrogen vary widely between species. Some grasses show no effects of fumigation (0.06 ppmv for 20 weeks) whilst in others as much as a 55 % reduction in dry weight have been recorded. The amount of damage suffered by a plant varies in its severity according to various factors such as concentration and length of exposure, plant age, light intensity, humidity, temperature and season (see previous sections). Symptoms are often divided into 'invisible' and 'visible' injury (see Ch. 1). The former where there is an overall reduction in growth but no obvious signs of damage is often associated with decreases in photosynthesis and transpiration.

During exposure to typical atmospheric levels of oxides of nitrogen, instances of visible injury are very rare. When it does appear it is often confused with visible damage caused by sulphur dioxide. Characteristic bleached (chlorotic) areas on the leaves associated with the damaged (necrotic) areas have been noted on a number of species exposed to high concentrations of nitrogen dioxide (e.g. 2.5 ppmv for 8 hours). Collapsed and dying bleached tissue occurs mostly at the apex of the leaves and along the margins, most severely on older leaves.

At more normal levels of atmospheric levels of oxides of nitrogen, damage is usually invisible, although ultrastructural studies with the electron microscope have revealed reversible swelling of the chloroplast thylakoid membranes similar to that caused by sulphur dioxide pollution. One visible feature often associated with atmospheric levels of oxides of nitrogen is quite noticeable but can hardly be described as damage because the polluted plants by comparison to those grown in clean air are often greener and as a consequence appear to be 'healthier' by virtue of the presence of oxides of nitrogen in the atmosphere. This flush is often only observed in the early stages of growth when exposure has been short and very often disappears after some time, to be replaced by quite obvious reductions in growth.

There have also been reports of beneficial effects of oxides of nitrogen upon plant growth. Out of six broad-leaved trees exposed to 0.06 ppmv nitrogen dioxide for 150 days, only silver birch showed significant increases in growth with exposure. Similarly, 28-days exposure of seedlings of American sycamore to levels of 0.1 ppmv oxides of nitrogen showed no significant effects on growth for the duration of the exposure but polluted plants grew significantly better than controls after the exposure was stopped. In conclusion, it appears that the response of plants to oxides of nitrogen pollution is species-dependent (or even variety or cultivar-dependent) as well as being dose-dependent and strongly influenced by the availability of nutrients such as nitrogen.

(d) Crops under glass

Although there is only a low risk of oxides of nitrogen entering a glasshouse from the outside atmosphere, internally-generated pollution is a real problem. It is frequent commercial practice in the horticultural industry to burn hydrocarbon fuels such as propane, natural gas or kerosene to enrich the atmosphere inside the glasshouse with carbon dioxide as well as to provide heat (see Sect. 6.2). Optimum benefits of carbon dioxide enrichment have been achieved by growers when levels are raised to over three times normal atmospheric levels (1000 ppmv). Since the cost of fuels is an important element in producing a crop commercially under glass, there has been an increasing tendency for growers to vent the flue gases from the hydrocarbon burners directly into the glasshouses. The burners generate mainly nitric oxide but this is not rapidly converted to nitrogen dioxide (Reaction 3.19) because ozone levels are much lower inside than outside the glass. In fact levels of oxides of nitrogen in heated glasshouses with carbon dioxide enrichment are comparable to those in the heaviest polluted out-of-door situations such as to be found in narrow streets with continuous flows of heavy traffic and flanked by high buildings. Some flueless heaters made especially for glasshouses can result in very high levels (> 1 ppmv) indeed.

Amounts of nitrogen released by combustion are significant to the health of plant and grower alike although the potential to provide for the nitrogen needs of a crop are surprising. A flueless heater is capable of releasing 100 kg nitrogen per hectare in a 100-day growing season – as much nitrogen as some crops under glass need without adding fertiliser. Certainly, tomato plants growing under low nitrogen can benefit from the extra atmospheric nitrogen and show stimulated growth. It would appear that modern cultivars of tomato plants which have been selected for their ability to succeed under commercial practice (involving high levels of nitric oxide) may have been inadvertently selected so that they have an enhanced tolerance to oxides of nitrogen pollution (see Sect. 7.3b).

Glasshouse pollution differs in three ways from equivalent outdoor conditions, all of which mitigate damage caused by the high levels of oxides of nitrogen. Firstly, there is very little air movement inside a glasshouse and therefore the boundary layers of air above and below the leaves are undisturbed. This means that much higher concentrations of oxides of nitrogen are required to give rates of uptake similar to those to be found in the turbulent air outside. Secondly, there is proportionally more nitric oxide than nitrogen dioxide inside glasshouses which may reduce the uptake rates because of the relative differences in solubility between nitric oxide and nitrogen dioxide in extracellular fluids once inside the plant. Finally, there is rarely a mixed pollutant situation in glasshouses although unburnt hydrocarbons may be an occasional problem. Fuels used for burning are almost sulphur-free so sources of sulphur dioxide are limited to what leaks in from outside. As Ch. 7 shows, avoidance of the combination of oxides of nitrogen, sulphur dioxide and ozone may be a critical factor in limiting the extent of plant damage.

Reductions in crop growth due to presence of high levels of oxides of nitrogen can be highly significant in terms of commercial economics. As much as the whole benefit that could be expected from carbon dioxide enrichment (see Ch. 6) may be wiped out. Tomato growth is depressed strongly by 0.2 ppmv nitrogen dioxide or 0.4 ppmv nitric oxide. Leaf distortion may also occur and the green coloration intensified. Fruit production is correspondingly reduced in quality and weight. Under conditions of low soil nitrogen the presence of low levels oxides of nitrogen may be beneficial to leaf area and plant weight. The situation is complicated because each cultivar of each species may give a different response so a balance between loss and gain is very difficult to predict or to achieve.

The metabolic response of the plants may also differ. Most species, like tomato, alleviate toxicity by extensive nitrogen metabolism. In the presence of nitric oxide, for example, most of the earliest commercial cultivars appear to increase the levels of their nitrite reductase activities or decrease those of nitrate reductase (see Fig. 3.9). This implies that the major input from the nitric oxide in the glasshouse atmosphere enters the plant as nitrite rather than nitrate. Equivalent exposure to nitrogen dioxide on the other hand enhances both enzymes signifying that nitrate rather than nitrite is more prevalent. When more modern commercial cultivars are examined under similar conditions the responses seen in the two enzymes are not so obvious and are not significant. This has the implication that, although natural responses in changes in metabolism can be utilised as a partial detoxification of the atmospheric oxides of nitrogen, plants that tolerate oxides of nitrogen do not have to resort to metabolic readjustments. They must have genetic tolerance by virtue of differences in uptake mechanisms rather than having increased capacity to deal with the products once they have entered.

Pepper plants behave quite differently to tomato plants in the presence of oxides of nitrogen. Instead of showing increases in the levels of nitrate reductase activity during exposure to nitrogen dioxide they sharply reduce the levels of enzyme in response to reductions of nitrite. The implication of this is that some uptake or transport mechanism has been restricted as a consequence of the additional source of nitrogen. How this operates is not known but it may be similar to the altered uptake characteristics shown by tolerant tomato cultivars.

(e) Prospectives

The fact that excessive levels of oxides of nitrogen in the atmosphere can damage vegetation is no longer in doubt. Nevertheless, how this occurs within the plant is only partially understood. By utilising normal pathways of nitrogen metabolism and natural means of adaptation such as the induction of the levels of either nitrate or nitrite reductase activities (Fig. 3.9), most plants attempt to detoxify and utilise the additional nitrogen from atmospheric oxides of nitrogen. In the longer term these adjustments are often inadequate. The net costs to the plant in energetic terms of general repair and the maintenance of detoxification processes are reflected in reduced growth and productivity.

What can be done to mitigate the effects of oxides of nitrogen? Clearly legislation to control or limit the amount of emissions from combustion processes is desirable but this must be hand-in-hand with technological improvement to burner design, manipulation of temperature and air/fuel ratios during combustion and more efficient exhaust trapping of pollutants.

There is another avenue of amelioration which undoubtedly is taking place naturally but could be assisted by plant scientists especially in circumstances where a significant loss of productivity and amenity in certain species is already taking place. Vegetation has a vast genetic potential for tolerance against environmental stresses. After light and water, one of the major limitations to plant growth is an adequate nitrogen supply. A wide range of inherent adaptations and homeostatic readjustments take place in most plants to ensure maximum utilisation of nitrogen under given circumstances. Some adapt better than others and survive when those less able to adjust are eliminated. Resistance or tolerance to pollutants is a well known phenomenon seen especially in studies of sulphur dioxide sensitivity. Clones of grasses and trees have been available for research study for some time although plants with such characteristics are rarely used in commercial breeding programmes. Plants that have enhanced sensitivity to pollutants can be employed for their diagnostic value as plant indicators in the field (see Sect. 1.4). For example, tobacco cultivar BEL W3 is especially sensitive to ozone and has been used to follow episodes of possible ozone injury in a number of countries.

Tolerance within plants towards oxides of nitrogen has rarely been demonstrated simply because it has never been looked for. Clearly a plant that can tolerate oxides of nitrogen and actually use the nitrogen as a nutrient is highly desirable. From studies on tomatoes it is apparent that those modern varieties that crop well in carbon-dioxide-enriched propane-heated glasshouses have enhanced tolerance to oxides of nitrogen, although this was not the trait intentionally followed by the tomato breeders whilst producing such cultivars. Perhaps the most significant work on oxides of nitrogen tolerance has been done at the University of Lancaster by Whitmore and Mansfield who fumigated a cultivar of grass (*Poa pratensis* var. Monopoly, currently recommended for resowing of UK pastures) for 8 months with low (60 ppbv) levels of nitrogen dioxide as well as with similar levels of sulphur dioxide and mixtures of the two. By the end of this period the unfumigated grass was at the stage of producing seed heads when haymaking normally takes place. They then ranked all the individuals of each treatment and the results are shown in Fig. 7.5. It is clearly apparent that within a single sowing of one cultivar a large variation still exists for tolerance or sensitivity towards nitrogen dioxide. Some NO_2-polluted individuals out-perform the best of the controls, others hardly grow. By selecting out these 'best' performers then growing their seed on and then cloning or micro-propagating their offspring, these may be further screened for evidence of desirable resistant characteristics. By doing this repeatedly it may be possible to find individuals that are able to use pollutants as nutrients and at the same time ignore their undesirable toxic side-effects.

3.3 HEALTH EFFECTS

(a) Industrial accidents

Threshold Limit Values (or OELs) set for industrial exposure of humans to oxides of nitrogen are 25 ppmv nitric oxide or 2 ppmv nitrogen dioxide lasting for 8 hours at a time (Table 1.4). These are well above what would normally be experienced by any member of the public but some unfortunate individuals occasionally experience exposures to massive concentrations of nitrous fumes. Unlike sulphuric acid gases which cause a violent breathing reflex to serve as a warning, nitrous fumes are more dangerous because a fatal quantity may be inhaled without immediate suspicion. Those industries involved in the nitration of various aromatics to form nitrocellulose (the basis of lacquers, films and celluloid) or nitrophenols used in the drug and dye industries, are prone to the production of nitrous fumes. Other industrial operations such as metal etching, photoengraving, welding in confined spaces or underground blasting can, under certain circumstances, also produce toxic and fatal releases of nitrous fumes. Fires, especially those involving plastics, produce vast quantities of nitrous fumes along with other irritants and hazardous vapours.

Even agricultural operations involving silage or stored hay can cause a condition known as 'silo-filler's disease' in the workers nearby. This occurs when excess fertilisers encourage anaerobic fermentation or denitrification which may then cause nitrous fumes to be given off. Lack of warning symptoms means that only after from 1 to 24 hours do exposure symptoms such as coughing, headaches and chest tightness develop. If particularly severe they are followed by either, sudden circulatory collapse or, congestion and water accumulation in the lungs (pulmonary oedema) about 36 hours after exposure. Once initial recovery is made then those individuals are likely to have a recurrent history of chronic chest complaints. Treatment with rest, oxygen, steroids and antibiotics is often recommended.

The damage is not just confined to the lungs. Nitric oxide may cause the haem of red blood cells to form a type of methaemoglobin by chelation mechanisms (see Sect. 3.3d) and excess blood nitrate may cause reduced blood pressure (i.e. acts as a vasodepressant). This, in turn, causes enhanced destruction of blood cells, liver and kidney defects (lesions) and increased efforts are made by the body to eliminate blood breakdown products from the blood into the urine as well as the bile duct. Because of deterioration caused to the liver and kidney as a consequence this often gives rise to jaundice and other related conditions. In some workers such as welders or silage workers who do not experience sudden massive doses (acute episodes) but have steady inhalations over a period of time, the chronic illness (silo-filler's disease) they later suffer is often unrecognised and not directly linked with their occupation.

The World Health Organisation has recently recommended to member countries of the United Nations a guideline of 0.21 ppmv of nitrogen dioxide not to be exceeded for a period of 1 hour. Over a period of 24 hours this should be reduced below 0.08 ppmv in order to safeguard human health.

(b) The domestic scene

Humans in modern society spend at least an average of 90 % of their time indoors but it is only recently that more attention has been paid to the possibilities of indoor pollution. Attempts to promote conservation of energy have concentrated on restricting indoor–outdoor changes of air which has made the problem much worse. Restricted combustion and poor oxidation rates mean that indoor levels of nitric

Table 3.3   Range of nitrogen dioxide concentrations detected in different rooms of domestic homes[†]

Concentration (ppbv)	Bedrooms (%)	Kitchens (%)	Living rooms (%)
0–21	48	11	40
22–42	33.5	29	49
43–63	13.5	19	18
64–84	2.5	13	5
85–105	2	10	2
106–126	0	7	1
127–147	0.5	3	1
148–168	0	1	1
>169	0	5	0

[†] Recalculated from Dutch surveys carried out between 1980 and 1982, see Schneider and Grant (1982) and references therein.

oxide predominate over those of nitrogen dioxide rather like the glasshouse situation (Sect. 3.2d). Levels of nitrogen dioxide around the home during the winter months can be considerable, especially in the kitchen and living rooms (Table 3.3).

It has always been thought that nitrogen dioxide is more toxic to human health than nitric oxide so monitoring studies have concentrated upon nitrogen dioxide to the complete neglect of nitric oxide even though levels of the latter indoors are known to be much higher by a factor of 2 or 3 at least. This neglect may only be justified on a short-term basis. By concentrating almost wholly on the effects of nitrogen dioxide as symptomatic of oxides of nitrogen as a whole, animal scientists may have already introduced an extra assumption which is then compounded by the fact that people in developed countries spend long period of their lives (90 %) indoors in atmospheres which often contain more nitric oxide than nitrogen dioxide. Experience with crops in glasshouses which have similar ratios of nitric oxide to nitrogen dioxide to those in homes would imply that account should be taken of this extra nitric oxide especially in long-term studies. One recent study of the effect of nitric oxide at a level of 1 ppmv, which is often reached as peak values in busy streets, on healthy people for just 2 hours has shown an increase in airway resistance thus demonstrating that complete neglect is not justified.

Levels of nitrogen dioxide indoors may range from 20 ppbv to over 250 ppbv to which should then be added the extra amounts of nitric oxide. On an industrial scale these higher levels of nitrogen dioxide are around one-eighth of the industrial Threshold Limit Value (or OEL) but experienced for far longer than a working day in industry by those who do not venture out quite as frequently, i.e. young children, mothers and the old, groups most at risk from a general medical consideration anyway. On top of this there is the added complication of smoking. Cigarette smoke contains oxides of nitrogen as well as a multitude of other potential irritants. Some samples of tobacco smoke may contain between 18–120 ppmv nitric oxide and similar amounts of nitrogen dioxide. Studies of households consisting of smokers as against non-smokers, or urban families cooking with gas rather than electricity, have shown higher prevalence of respiratory symptoms in the children of the smoking and gas-cooking households. Only in the latter epidemiological studies could the medical problems be specifically associated with enhanced levels of oxides of nitrogen because there are so many other harmful agents in cigarette smoke which make definitive interpretations much more difficult.

In such epidemiological surveys it is difficult to determine if oxides of nitrogen are the cause of respiratory infections or diminished lung capabilities, or whether they act in concert with other agents, such as sulphur dioxide, particulates or photochemical oxidants. Large-scale surveys of lung (pulmonary) function of workers, their children and their parents, some of which were exposed to daily averages of up to 160 ppbv of nitrogen dioxide in the presence of other pollutants, showed no clear associations of atmospheric level of nitrogen dioxide and pulmonary function. On the other hand, children and housewives in homes cooking with gas, where hourly peaks of 0.27–1 ppmv were experienced, showed a positive increase in respiratory illness whilst epidemiological studies on other groups where hourly peaks were much lower (0.08 ppmv) showed no such increased risk.

Healthy and bronchitic volunteers have been exposed to atmospheres containing only nitrogen dioxide as well as other pollutants, singly and in combination for up to 2 hours. Only levels above 1.6 ppmv caused increases in airway resistance or depressions in lung capacity but no difference between the healthy and bronchitic volunteers were found over this short period. Certain asthmatic patients on the other hand show marked increases in airway resistance to much lower exposures (0.1 ppmv) of nitrogen dioxide. It would appear that these individuals are predisposed to react to a wide range of irritants that cause the release of local hormones which cause constriction of blood vessels, reduced lung function and release of mucus. Oxides of nitrogen appear to act as one such causative agent to accelerate the onset of respiratory problems in those individuals already predisposed to asthma.

(c) Lung damage

The major difference between human studies of the effects of a pollutant gas and those on plants, microbes and other animals, is that there is rarely the possibility of comparable and equivalent biochemical and physiological investigations. Only the pathological consequences may be directly investigated. Animal studies, on the other hand, can only be evaluated and extrapolated with some degree of error as to the hazards presented by similar pollutant gases to man, although tissue cultures of human cells or biopsy material is of use in certain circumstances. A vast amount of research has been carried out on a wide range of animal studies designed to model the human situation and assess the consequencs of short-term and long-term exposures of man to oxides of nitrogen. Wide differences exist between models and conditions of exposure but some salient features do emerge.

Penetration and retention of nitrogen dioxide depends on many factors. In tissue fluids, for example, their reactivity, their concentration in the air taken in, the length of exposure, or the depth and frequency of breathing are all important factors. As nitrogen dioxide is quite soluble by virtue of Reaction 3.18 it will readily enter the tissue fluid (mainly as nitrate) very like it does into the extracellular fluid of plant tissues (Sect. 3.2a). By the same measure nitric oxide has the potential to enter with rather more difficulty as a mixture of nitrate, nitrite and undissociated nitrous acid. Consequently, airways to the lungs, passing through the nasal regions and into the terminal bronchioles of the lungs and finally into the gas-exchange areas (the alveolar parenchyma), all have their surfaces exposed to the products of oxides of nitrogen in solution. During light breathing about 40 % of the oxides of nitrogen are taken up in the nose and throat regions but, as exercise is taken and more breathing takes place through the mouth, this balance shifts towards lung tissues which become the most important uptake surfaces. At low concentrations, the most sensitive areas are the junctions between the bronchioles and the alveolar regions but ciliary cells and squamous surface cells may also be subjected to injury and loss. With higher concentrations the injury may extend to the larger airways and into the deeper tissues. Other changes include accumulation of tissue fluid (oedema), cellular debris or mucus, and an aggregation of certain white blood cells (macrophages) which accumulate as one of the lines of defence against an increased likelihood of infection. Animal lungs have a remarkable ability to rid themselves of inhaled microbes, some of which are pathogenic and are essential in order to maintain the sterility of their alveolar regions. Low concentrations of nitrogen dioxide actually disturb these natural defences within the lungs of animals and cause

Table 3.4 Likely effects of nitrogen dioxide upon humans

Pollutant range[†] (ppmv)	Symptoms
0 – 0.21	No effects
0.11 – 0.21	Slight odour detected
0.22 – 1.1	Some metabolic effects associated with either toxicity, adaptation or repair of lung tissues (e.g. inhibited metabolism of prostaglandin E_2)
1.1 – 2	Significant changes to respiratory rate and lung volume, enhanced susceptibility to infection and evidence of tissue repair
2.1 – 5.3	Deterioration of lung tissue (e.g. loss of cilia) not balanced by repair mechanisms
above 5.4	Gross distortion of lung tissues and emphysema, possible death if prolonged

[†] Over a period of 2 hours (for asthmatics, etc., these values are much lower).

them to become more susceptible to bacterial attack (see Table 3.4).

Over the longer term it would appear that some of the damage caused to the epithelial layers by pollutants may be repaired if the levels of nitrogen dioxide are low or reduced. Newer cells formed during repair are relatively resistant to the effects of the gas by comparison to those originally damaged. Other longer-term studies have shown that oxides of nitrogen-induced changes affect the blood circulation. The partial pressure of oxygen may be reduced and that of carbon dioxide, as well as the blood pH, may rise. The extent of change is a question of scale – the amount of pollutant, or the period of exposure and just how well animal studies extrapolate to human conditions. Table 3.4 is an attempt to cover most of the essential changes that have been detected across a range of longer-term animal studies which probably apply to humans. Unfortunately, the majority of animal studies have been undertaken at atmospheric levels of nitrogen dioxide way beyond what most people would experience even in the most polluted of homes, streets, factories or glasshouses, and their validity is questionable.

(d) Metabolic changes

The effects of oxides of nitrogen upon lung tissues tend to resemble those of ozone rather than those of sulphur dioxide or ammonia because the latter are scrubbed more efficiently from inhaled air by the nasal and throat passages on the way to the lungs. When they finally enter the surface fluid covering the lung alveoli, nitrate or nitrite (as well as undissociated nitrous acid and more hydrated forms of nitrogen dioxide or nitric oxide) may coexist (Sect. 3.2a and 3.2c) and gain access to the cells. It is quite likely that other forms of nitrogen could also be formed. For example, amino-acids and other amines could be converted to more active nitroso-derivatives (Reactions 3.41 and 3.42) by undissociated nitrous acid which is in equilibrium with nitrite. Owing to the fact that lung tissue has a poor ability to oxidise nitrite to nitrate or to reduce nitrite to ammonia means that nitrosamine production could be more of a hazard to animals than plants especially as most of them are confirmed carcinogens.

Lipids of the cell membrane and elsewhere in the cell could be susceptible to auto-oxidation induced by nitrogen dioxide or subjected to *cis*-to-*trans* isomerisations of their double bonds. Experimental simulations in non-aqueous media have demonstrated that at least four different free radical species are generated by attack of nitrogen dioxide upon unsaturated linkages in lipids but the extent to which they occur in semi-aqueous conditions is not known. The damage they cause to proteins containing the amino-acid tryptophane could have some importance whilst the lipid peroxidation (see Ch. 5) that may occur even after only short exposures (up to 1 ppmv) may persist for up to 48 hours after exposure. Intermittent rather than continuous exposures to nitrogen dioxide or deficiencies of vitamin E (α-tocopherol, a natural free radical scavenger synthesised by plants and taken up in the diet) make metabolic disturbance like these more likely. It has been suggested that auto-oxidation is promoted by nitroso free radicals which abstract the methylene hydrogen from between double bonds to give alkyl free radicals which then react with oxygen to give fatty acid hydroperoxides (see Ch. 5). Natural antioxidants like vitamin E then react with these peroxyl radicals rather than with nitrogen dioxide directly in order to suppress these chain reactions. However, nitrogen dioxide-catalysed peroxidation differs from ozone-induced peroxidation because the initial peroxyl step is much slower and more readily prevented by phenolic antioxidants such as vitamin E, peroxide decomposers like the haem proteins, or protective enzymes like glutathione peroxidase or reductase and disulphide reductase (see Ch. 5). Levels of these protective enzymes have been found to rise during exposure to nitrogen dioxide. This gives additional support for the idea that lung tissues attempt to counteract the tendency towards enhanced peroxidation.

Both nitrogen dioxide and nitric oxide may form hydroxyl radicals (Reactions 3.44 and 3.45) if they react with hydrogen peroxide which may be present as a consequence of smoking or the additional presence of ozone (Ch. 5). The hydroxyl radicals are so reactive they attack most other organic compounds (see Appendix, Sect. 8.2) but they specifically attack the α_1-antiproteinase inhibitor which normally prevents proteinases (enzymes such as elastase which hydrolyse specific proteins like elastin) from their destructive influence. In this instance, the hydroxyl radical appears to oxidise a critical methionine group in the α_1-antiproteinase. This being so then elastase hydrolyses the elastin in the lung tissues which reduces their elasticity and hence the effectiveness of contractions and expansions during breathing – very similar to the diminished lung capacity functions found in the disease known as 'emphysema'.

$$NO_2 + H_2O_2 \longrightarrow HONO_2 + OH^{\bullet} \tag{3.44}$$

$$NO + H_2O_2 \longrightarrow HONO + OH^{\bullet} \tag{3.45}$$

Changes in blood composition of humans are readily detected and have been of more use in detecting and monitoring effects that oxides of nitrogen have upon them. Both nitric oxide and nitrogen dioxide may complex with the iron of haemoglobin to form a type of methaemoglobin which prevents the binding of oxygen and has a characteristic electron spin resonance (ESR) spectrum. Unfortunately, there is no evidence that this type of methaemoglobin arises *in vivo* at the levels of oxides of nitrogen to be found even under some of the most adverse working conditions.

Of much more relevance are the changes in some of the regulatory non-respiratory effectors found in the capillary beds of lungs (Table 3.5 and Fig. 3.11). A number of these modulators or local hormones counterbalance each other and influence local blood pressure within lungs. By doing so they alter the ratio of blood perfused through the tissue relative to the amount of air ventilated through the lungs. Minute changes in blood pressure due to alterations in balance between these modulators have quite significant effects on physiological function of lung tissue because the lung is low-pressure–high-flow system. Some modulators or local hormones are made and most are broken down within lung tissue. Controls on the rates of their synthesis and degradation allow modulation of the levels of some of these effectors and thereby may exert an even finer control upon their action.

One of these modulators, prostaglandin E_2 has been studied during exposure of rats to low doses (0.2 ppm for 3 hours) of nitrogen dioxide. Normally this prostaglandin causes local constriction of blood vessels to increase blood pressure and is metabolised to inactive forms (first to 15-ketoprostaglandin, and thence to 13, 14-dehydro-15-ketoprostaglandin E_2) which are eliminated from the capillary cells by the blood flow. By virtue of this metabolism the constricting effect

Table 3.5 Effect on local blood pressure caused by changes in the levels of local modulators

Modulator	Lung blood pressure	
(local hormone)	Enhanced modulator levels†	Reduced modulator levels‡
Angiotensin II	Increased	Decreased
Bradykinin	Decreased	Increased
Prostaglandin E_2	Increased	Decreased
Prostaglandin I_2	Decreased	Increased
Serotonin (5-hydroxy-tryptamine)	Increased	Decreased

† Caused by increased rates of synthesis and/or uptake (in addition to reduced rates of breakdown or elimination) of local modulator.
‡ Caused by the converse of†.

of the prostaglandin E_2 is reduced and blood pressure falls. However, when exposure to nitrogen dioxide is experienced, the metabolism of prostaglandin E_2 to the 15-keto- and 13, 14-dehydro-15-keto-forms is inhibited and hence the vasoconstriction (increased blood pressure) is not removed. This effect persists for 60 hours after the cessation of exposure. The response is thus highly sensitive to nitrogen dioxide and, furthermore, it is accumulative. It could be one of the initiation

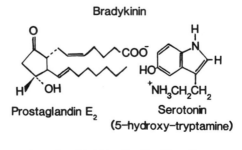

Arg–Pro–Pro–Gly–Phe–Ser–Pro–Phe–Arg

Bradykinin

Prostaglandin E_2 Serotonin
(5–hydroxy–tryptamine)

Asp–Arg–Val–Tyr–Ile–His–Pro–Phe

Angiotensin II

Fig. 3.11 Structures of a number of important local modulators of lung blood pressure in humans.

events in the development of some of the obstructive symptoms only observed at much higher levels of exposure. No doubt there are other metabolic disturbances of some of the other modulators involved in control of lung function as well. Modern methods of diagnostic detection of these effectors and their precursors or products should now enable this type of investigation to be translated to the human situation with some confidence of predictive value with respect to the hazards of ambient concentrations of oxides of nitrogen upon health.

SELECTED BIBLIOGRAPHY

Freeney, J. R. and Simpson, J. R. (eds.), *Gaseous Loss of Nitrogen from Plant–Soil Systems*. Martinus Nijhoff/Dr. W. Junke Pub., 1983, The Hague.

Grosjean, D., *Nitrogenous Air Pollutants: Chemical and Biological Implications*. Ann Arbor Sci. Publ., 1979, Ann Arbor, Mich.

Lee, J. A. and Stewart, G. R., Ecological aspects of nitrogen assimilation. *Advances in Botanical Research* 6, 1978, 1–43.

Ministry of Agriculture, Fisheries and Food, *Nitrogen and Soil Organic Matter*. Technical Bulletin no. 15. 1969.

National Academy of Sciences. *Nitrogen Oxides*. Committee of Medical and Biological Effects of Environmental Pollutants. 1977, Washington, D.C.

Rowland, A., Murray, A. J. S. and Wellburn, A. R., Oxides of nitrogen and their impact upon vegetation. *Reviews of Environmental Health 5*, 1985, 295–312.

Schneider, T. and Grant, L., *Air Pollution by Nitrogen Oxides*. Elsevier Sci. Publ. 1982, Amsterdam, Oxford and New York.

VDI-Kommision Reinhaltung der Luft, *Stickstoffoxide Bibliographie 1970–1974*. Dusseldorf.

Watkins, L. H., *Environmental Impact of Road and Traffic*. Applied Science Publishers, 1981, London and New Jersey.

4

Acid rain

'When your fountain is choked up and polluted,
   the stream will not run long,
   or will not run clear with us,
   or perhaps with any nation.'

Edmund Burke (1729–1797)

4.1 FORMATION AND DEPOSITION OF ACIDITY

(a) Definitions and distinctions

The term 'acid rain' was first used by Robert Angus Smith in 1872
to describe the acidic nature of the rain falling around Manchester
in one of his early reports as the first Chief Alkali Inspector of the
UK. The conception that 'acid rain' cannot be considered in isolation
from atmospheric pollution in general was introduced in Ch. 1 and
preceding chapters have shown that there are also gas-phase reactions which ultimately produce acidity upon the surfaces of the Earth.
These processes which do not involve the water droplets are known
as 'dry deposition'. Complementary processes which do involve
gas–water phase reactions in the atmosphere which then reach the
surface by precipitation are therefore more properly called 'wet
deposition'. Such overall processes would include acid-producing

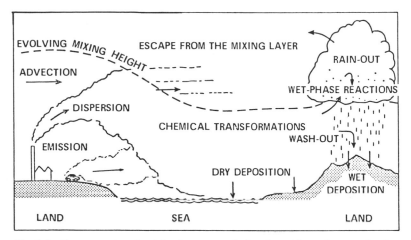

Fig. 4.1 Processes involved in the wet and dry deposition of atmospheric pollution. (Courtesy of the Watt Committee on Energy, UK.) By comparing Fig. 4.1 with Fig. 1.1 one may observe how different nations (probably subconsciously) perceive their relationship towards cause and effect. (Clue: look in the middle.)

mechanisms as well as non-acid-forming processes. Thus the popular phrase 'acid rain' as a process is only a part of wet deposition, albeit a very important and major component. Most of the events involved in the wet deposition of atmospheric pollution are shown in Fig. 4.1.

Wet deposition by mist, rain, hail, sleet or snow is often collectively called 'precipitation' or 'wet precipitation'. As this is a series of intermittent events, this means that the effects of acid rain on the ground are also experienced irregularly. Table 4.1 lists the different types of deposition and the terms usually associated with each event to make these definitions clearer.

Formation of droplets which collect and react with pollutants within clouds is a very efficient process known as 'rain-out' (Fig. 4.1). Removal rates by rain-out depend on a variety of different factors, especially the type and amount of pollutant or the mean size and temperature of the raindrops and snowflakes. Of lesser importance is the process of 'wash-out' beneath clouds although this is more important as a means of removing particulates (see Sect. 1.7).

Emissions of primary pollutants such as sulphur dioxide or nitric oxide have already been considered in Ch. 2 and 3. Subsequent gas-phase transformations in the plumes or as they escape from the mixing layer (Fig. 4.1) to react with light, oxygen, etc., have also been examined. However, it must be appreciated that most of these gas-phase

Table 4.1   Terms used to describe physical processes involved in
atmosphere to ground (or water) surface transfer

Deposition	Process	Terminology
Wet	Rain, snow, sleet or hail hitting a surface[†]	Precipitation
Wet	Fog and mist aerosols hitting a surface[†]	Wet impaction (or occult deposition[‡])
Dry	Solid particles (other than those above) a hitting surface[†]	Particulate deposition or sedimentation
Dry	Dry aerosols hitting surface[†]	Interception
Dry	Gases coming in contact with surfaces[†] with subsequent chemical or physical reactions thereafter	Gaseous deposition or uptake

[†] Of the Earth.
[‡] When mists cause condensation on surfaces.

reactions responsible for ultimately forming acidity outside clouds are
relatively slow. Four such gas-phase processes were, for example,
considered in Ch. 2 (Reactions 2.1 to 2.8). Of these four possibilities,
reaction of sulphur dioxide with hydroxyl radicals (formed from ozone
or monatomic oxygen and water) was the most likely (Reactions 2.7
and 2.8). Similar dry gas-phase mechanisms (e.g. Reaction 3.22) exist
for the formation of nitric acid from nitrogen dioxide but it is significant that the gas-phase oxidation of nitrogen dioxide by hydroxyl
radicals proceeds ten times faster than the equivalent reactions
involving sulphur dioxide.

Gas-phase reactions can also generate organic acids. For example,
reaction of peroxyl radicals $(OH_2^{\bullet})$ with aldehydes (HCHO,
CH_3CHO, etc.) may lead to the formation of formic acid, acetic acid
and higher acids. These organic acids are thought to contribute
between 5 % and 20 % towards the total acidity of tropospheric
atmospheres.

For a fuller explanation of acidity, and the relationship between
pH (i.e. the negative logarithm of the hydrogen ion concentration)
and microequivalents, please refer to the Appendix, Sect. 8.5.

(b)   Cloudwater acidity

The best starting point to consider the acid-forming reactions of
cloudwater is to ask the question: 'What would be the pH of cloud-

water if there was no atmospheric pollution?'. If one leaves aside the possible contributions of sulphur dioxide and oxides of nitrogen from volcanoes, swamps and lightning strikes, etc., then the major 'natural' gas in the atmosphere to contribute acidity would undoubtedly be carbon dioxide (see also Sect. 6.2).

The aqueous reactions of carbon dioxide are as follows (Reactions 4.1 to 4.3):

$$CO_2 \text{ gas} + H_2O \rightleftharpoons H_2CO_3 \tag{4.1}$$

$$H_2CO_3 \rightleftharpoons HCO_3^- + H^+ \tag{4.2}$$

$$HCO_3^- \rightleftharpoons CO_3^{2-} + H^+ \tag{4.3}$$

As the pK_a of Reaction 4.3 is as high as 10.3 (i.e. when concentrations of carbonate and bicarbonate are equal) then Reaction 4.2 has by far the greatest influence upon the acidity of natural atmospheric systems. Furthermore, the amount of gas in solution is a function of the partial pressure of that gas in atmospheres, which in turn is proportional to the mole fraction (i.e. the number of moles of a component divided by the total number of moles in the mixture) of that gas. In the case of carbon dioxide this means that if the partial pressure of atmospheric carbon dioxide is 0.00035 atm, then Henry's law constant (K_H) is as follows:

$$K_H = \frac{[H_2CO_3]}{[CO_2 \text{ gas}]} = 3.79 \times 10^{-2} \text{ mol l}^{-1} \text{ atm}^{-1}$$

and the equilibrium constant ($K_{4.2}$) of Reaction 4.2 is given by:

$$K_{4.2} = \frac{[H^+] [HCO_3^-]}{[H_2CO_3]} = 4.5 \times 10^{-7} \text{ mol l}^{-1}$$

By combining and rearranging these two expressions, one achieves the following equation:

$$[HCO_3^-] = \frac{[CO_2 \text{ gas}] \times K_H \times K_{4.2}}{[H^+]}$$

Thus, if the concentration of bicarbonate in pure water is equal to the hydrogen ion concentration, then by substitution one arrives at the following:

$$\begin{aligned} [H^+]^2 &= [CO_2 \text{ gas}] \times K_H \times K_{4.2} \\ &= 0.00035 \times 3.79 \times 10^{-2} \times 4.5 \times 10^{-7} \\ &= 5.97 \times 10^{-12} \text{ mol}^2 \text{ l}^{-2} \end{aligned}$$

therefore $\quad [H^+] = 2.44 \times 10^{-6} \text{ mol l}^{-1}$
and hence $\quad pH = -\log [H^+] = 5.61$

This means that, however clean the atmosphere and disregarding any wind-blow of alkaline materials, natural cloudwater will always be slightly acidic.

Exactly the same approach may be used to calculate the effect of acidic atmospheric gases like sulphur dioxide and nitrogen dioxide, even though they dissolve and dissociate much more readily than carbon dioxide, because their respective Henry's law constants and relevant dissociation constants are much greater. Of the various equations for sulphur dioxide (Reactions 4.4 to 4.6), only the first two are relevant to cloudwater acidity because the pK_a of Reaction 4.6 is above neutrality (pK_a 7.2) so, at pH values less than 5.6, very little sulphite would exist even if it were not oxidised quickly.

$$SO_2 \text{ gas} + H_2O \quad \rightleftharpoons \quad H_2SO_3 \qquad (4.4)$$

$$H_2SO_3 \quad \rightleftharpoons \quad H^+ + HSO_3^- \qquad (4.5)$$

$$HSO_3^- \quad \rightleftharpoons \quad H^+ + SO_3^{2-} \qquad (4.6)$$

If the K_H for Reaction 4.4 is 1.24 mol l^{-1} atm^{-1} and the equilibrium constant for Reaction 4.5 is 1.27×10^{-3} mol l^{-1} then an atmosphere of 100 ppbv sulphur dioxide would theoretically produce acidity, due only to bisulphite formation, equivalent to pH 4.9 in clouds nearby if calculations very similar to those for carbon dioxide above are followed. However, 100 ppbv sulphur dioxide is equivalent to a moderately polluted urban atmosphere. Other mechanisms beyond the simple dissolution of sulphur dioxide must therefore be involved in order to form acid rain. This acidification by oxidation is covered in the next section.

Calculations have been also done to assess the natural contribution of the global sulphur cycle (see Ch. 2) to cloudwater acidity without the added man-induced emissions and leaving aside the complications of alkalinity arising from ammonia, limestone particulates, wind-blow, etc. These show that, in pristine conditions, cloudwater would naturally range between pH 4.5 to 5.6 because of seasonal and local variations. It follows that any consideration of the mechanisms of oxidation of bisulphite compounds and the development of acid rain in cloudwater should therefore only include those ranges of pH-dependent reactions which occur below pH 5.6.

(c) Generation of sulphuric acid

If bisulphite is oxidised into bisulphate within water droplets then the whole relationship between the uptake of sulphur dioxide and the ionisation of sulphurous acid (Reactions 4.4 and 4.5) is disturbed. All the relationships are altered in favour of more sulphur dioxide equilibrating with cloudwater and thus more acidity will be formed.

In addition to the gas-phase reactions involving hydroxyl radicals that produce acidity (see Ch. 2), oxidation mechanisms within cloud droplets are important acid forming events. But many oxidising agents formed in the atmosphere such as ozone, hydrogen peroxide (H_2O_2), peroxyacylnitrates (PANs, e.g. $CH_3COO_2^{\cdot}NO_2$, see Ch. 5) or other free radicals (e.g. $CH_3O_2^{\cdot}H$ and $CH_3COO_2^{\cdot}H$) do not actually react with gaseous sulphur dioxide to any great extent. However, when they are dissolved in cloudwater, they readily oxidise dissolved bisulphite ions (e.g. Reactions 4.7 and 4.8) which then form sulphate anions and therefore more acidity (Reaction 4.9).

$$O_3 + HSO_3^- \longrightarrow HSO_4^- + O_2 \qquad (4.7)$$

$$H_2O_2 + HSO_3^- \longrightarrow HSO_4^- + H_2O \qquad (4.8)$$

$$HSO_4^- \rightleftharpoons H^+ + SO_4^{2-} \qquad (4.9)$$

$$2HO_2^{\cdot} + M \longrightarrow H_2O_2 + O_2 + M \qquad (4.10)$$

Hydrogen peroxide, like ozone (see Ch. 5), arises mainly from gas phase by a series of reactions (e.g. Reaction 4.10) involving a number of atmospheric free radicals (see Appendix, Sect. 8.2). However, some of these hydrogen peroxide-forming radicals (e.g. peroxyl, $HO_2^{\cdot}$) can also occur in cloudwater droplets and form significant amounts of hydrogen peroxide.

The most important distinction between Reactions 4.7 and 4.8 is that oxidation by ozone, although rapid at pH 5, is pH-dependent and falls off quickly with increasing acidity along with the solubility of sulphur dioxide. Hydrogen peroxide oxidation (Reaction 4.8) on the other hand is virtually independent of pH which means that at pH 5 ozone may be equally or more important as a means of forming bisulphite in water droplets. Once acidity increases then hydrogen peroxide oxidation becomes much more important. Most of the other free radicals (e.g. PANs) behave like ozone in this respect although the peroxyacetic acid ($CH_3COO_2^{\cdot}H$) and methylhydroperoxide ($CH_3OO_2^{\cdot}H$) radicals more closely resemble the behaviour of hydrogen peroxide. Thus ozone and PAN-dependent oxidations, as well as the pH-dependent Fe^{3+}- or Mn^{2+}-catalysed oxidation of bisulphite by oxygen, fall rapidly as the acidity of cloudwater increases. Alternative oxidations such as those caused by nitrite or those catalysed by carbonaceous particles are also relatively unimportant during most tropospheric conditions. This means that levels of cloudwater hydrogen peroxide are the major and limiting controls upon acidity production under a wide variety of climatic conditions. To support this contention significant levels of hydrogen peroxide have been detected in clouds by instruments carried by aircraft flying in a variety of conditions. Average levels of 0.1 ppbv hydrogen peroxide, with peaks up to 0.9 ppbv, have been recorded by such means and these correlate quite well with samples collected on mountain tops.

(d) Nitric acid formation

As Ch. 3 has already shown, the dominant sequence of reactions for the formation of nitric acid (Reactions 4.11 to 4.14) in colder conditions takes place both in the gas phase (Reactions 4.11 and 4.12) and as soon as nitrogen peroxide comes in contact with water (Reactions 4.13 and 4.14).

$$O_3 \; + \; NO_2 \longrightarrow \; NO_3 + O_2 \tag{4.11}$$

$$NO_3 \; + \; NO_2 \longrightarrow \; N_2O_5 \tag{4.12}$$

$$N_2O_5 \; + \; H_2O \longrightarrow \; 2HONO_2 \tag{4.13}$$

$$HONO_2 \qquad \rightleftharpoons \; H^+ + NO_3^- \tag{4.14}$$

The alternative reaction of nitrogen dioxide with hydroxyl radicals to form nitric acid (Reaction 4.15) wholly in the gas phase is essentially a warm summer reaction. This transition from one gas-phase reaction to a series of gas–liquid-phase reactions, either from warm to cold, or, from summer to winter conditions, offers an explanation why nitrate is deposited more evenly over the seasons than sulphate. The analogous gas-phase reaction to Reaction 4.11 for sulphur dioxide (i.e. Reaction 4.16) is very slow and therefore the reaction with hydroxyl radicals (Reactions 2.7 and 2.8), which would normally carry out the sulphur dioxide equivalent of Reaction 4.15, ensures that this is a warm and strong-light phenomenon which causes most sulphate to be formed and deposited in late summer and early autumn.

$$OH^{\bullet} + \; NO_2 + M \longrightarrow \; HONO_2 + M \tag{4.15}$$

$$O_3 \; + \; SO_2 \qquad \longrightarrow \; SO_3 + O_2 \tag{4.16}$$

(e) Other sources of acidity and alkalinity

Analysis of the ionic balance in rainwater shows clear differences between samples that fall on the land and those on the oceans. Table 4.2 shows the order of importance of different ionic species that contribute both to global acidity and alkalinity. Sodium and chloride ions are the dominant ions of seawater and are injected into the atmosphere by an active upward mechanical process known as 'bubble-bursting' and are carried inland as wind-blown spray. On land, natural injection of alkaline substances such as calcium, magnesium or potassium ions into the atmosphere as wind-blow from

Table 4.2 Order of importance for different ionic species contributing to global acidic and alkaline deposition

Surface	Acidity	Alkalinity
Over land[†]	$SO_4^{2-} > NO_3^- >> Cl^-$	$Ca^{2+} > Mg^{2+} > NH_4^+ > K^+ > Na^+$
Over oceans[†]	$Cl^- > SO_4^{2-} >> NO_3^-$	$Na^+ > Mg^{2+} > Ca^{2+} > K^+ >> NH_4^+$

[†] At distances beyond 100 km from the other.

natural soils far outweigh man-made emissions of particulates such as fly-ash from combustion or dust from mining operations and cement production. The largest single man-induced source of alkaline emission is ammonia caused by enhanced agricultural operations (see Ch. 3). Loadings in the atmosphere are highly variable, depending on wind speed, the frequency or duration of precipitation and the nature of the vegetation beneath. Ammonium levels in rainwater are highest during drought or at spring-time or autumn-time when ploughing operations take place. The total mass flux of alkaline substances in cereal-producing prairie regions is often as great if not greater than the mass flux of sulphur caused by emissions from highly industrialised regions (i.e. more than 50 kg ha^{-1} y^{-1}). This means on occasions that some of the rain which falls in the prairie areas of North America can be distinctly alkaline (i.e. above pH 8).

(f) Dispersion and transport

Natural and man-made emissions are released into the atmosphere by a complicated mixture of physical processes. Apart from volcanoes, most natural and some man-made emissions (e.g. by traffic) originate at ground level whilst others are ejected from stacks often over a hundred metres high. All emissions reach an altitude dependent on the mixing characteristics of the atmosphere prevailing at the time of release. Sometimes at night, and all day over snow-covered ground, the atmosphere is stable with little vertical mixing. Under these conditions stack emissions once they have reached a balanced height with temperature then spread out slowly and travel quite long distances as a coherent plume. Local topography plays a considerable part. If stacks are not high enough, emissions are trapped in valleys and flow along water-courses in the manner of liquids. During the day, heating of surfaces and lower layers of air causes convection which may vertically lift emissions to considerable heights. Plumes may also become unstable and show the formation of eddies and loops.

These are often the cause of intense local concentrations of pollution if unstable looping causes the plume to dip as far as ground level. Again, local topography plays a part. High ground and tall buildings, for example, can cause downdraughts which achieve high local pollutant concentrations over quite small areas.

Pollutants at any height may be carried away (or advected) by winds and therefore climatic factors determine the direction and speed of transport. At the same time dispersion may take place by turbulent eddying. However, there are also weather conditions that cause the accumulation of pollutants. Slow-moving, high-pressure areas with low advection and dispersion rates, for example, have the effect of concentrating pollutants. When these eventually meet an advancing low-pressure system, an episode of highly acidic rain may be experienced often a thousand kilometres or so from the original sources. Such high–low pressure transitions are frequent over the eastern USA and northern Europe.

(g) Removal and deposition

As has already been mentioned, falling rain, snow, sleet, etc., are precipitation processes that transfer pollutants and their acidic products on to the surface of the land or the oceans by an overall process known as 'wet deposition'. Because this precipitation is nearly always acidic (i.e. with a pH less than 5.6) it is more frequently called 'acid rain'. The mechanism of wet deposition differs considerably from that of dry deposition because the rates of dry deposition (D_d) are directly dependent on pollutant concentration velocity (C_a) and deposition velocity (V_d) which, in turn, depend on the nature of the uptake or receiving surfaces on the land or the sea (see Ch. 3). By contrast, rates of wet deposition (D_w) do not depend on the underlying surface characteristics but on the precipitation rate (P), the wash-out ratio (W, i.e. the concentration of dissolved pollutant per unit mass of cloudwater or rain divided by the concentration of the same pollutant or precursor per unit mass of air), and the ambient air concentration (C_a). Values of W may range from tens to several thousand.

$$D_d = C_a V_d$$

$$D_w = WPC_a$$

Large networks of rainfall monitors exist and it is now possible to project the extent of acid rain onto a global scale. Figure 4.2 shows one of the projections for precipitated acidity over the northern hemisphere during the late 1970s. The pH range of values over the major portions of northern Asia and America vary between pH 5 and pH 5.5. As Sect. 4.1(b) has already shown, pristine conditions (i.e.

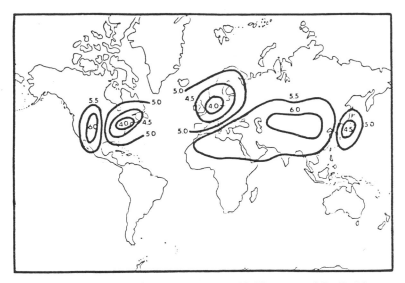

Fig. 4.2 Global ranges of mean rainwater pH (Courtesy of Dr D. M. Whelpdale and the WHO Centre on Surface and Ground Water, Canada and also Dr G. Gravenhorst and the Plenum Press).

in the absence of man-made pollution) would produce natural acidity in rain around pH 5.6. Areas regularly receiving over ten times this amount of acidity (less than pH 4.5) or greater are those regions along the eastern seaboard of the USA or northern Europe. The greatest annual acidity (less than pH 4) in rain falls on Denmark, southern Sweden and the north of New York State and amounts to nearly forty times background levels. Deposition patterns for sulphate ions are similar to those of pH and in the most heavily affected regions may reach 2.5 g of sulphur per square metre per year, with nitrate deposition rates an additional one-third of these values. Indeed, closer to sources and downwind from them, nitrate deposition rates can sometimes equal or exceed those of sulphate, especially in northern Europe.

Analysis of rainfall data from the 120 sites of the European Atmospheric Chemistry Network with 5 or more years of continuous results has shown that 29 of them show a significant trend of increasing acidity; with 23 showing increased deposition of sulphate and 55 having more nitrate (Note, 5 had less acidity and 1 had less sulphate). This tendency towards increased acidity was particularly abrupt after the year 1965. It was noticeable that the numbers of individual incidences of rainfall with exceptionally low pH values have tended to increase since 1965 although instances of rainfall with pH values lower than 3.4 are rarely observed. This is believed by some to be a lower limit because lifetimes of cloud or rainwater droplets are limited and at such low pH values the solubility of sulphur dioxide is restricted. Moreover, rates of oxidation of sulphur dioxide are much reduced at low pH values because the reaction with ozone is pH-dependent.

However, suspensions of acidic droplets in mists may have much lower acidity than pH 3.4 especially at high altitudes because physical changes such as evaporation (see Sect. 4.2b) take place between the atmosphere and these small water droplets (1–100 μm) which also pick up extra oxidants. These highly acidic mists (< pH 2.75) drift around trees at high altitude and adhere to needles by a process known as 'droplet interception' or 'occult deposition'. How much harm they subsequently cause has not yet been fully established (see Sect. 4.2b).

4.2 CONSEQUENCES TO PHYSICAL SYSTEMS

(a) Plant nutrients and soil acidity

Both wet- and dry-deposition processes transfer sulphates and nitrates to soil systems but deposition of acid rain also has the effect of increasing soil acidity and cause the mobilisation or leaching of nutrient cations which then threatens soil fertility. Forest vegetation also tends to remove pollutants from polluted atmospheres, not only by direct uptake (see Chs. 2 and 3), but also to increase the adsorption of ions so that later precipitation or throughfall (stemflow when it runs down the trunk) causes an enrichment of certain elements at the bases of trees. This is especially the case with sulphate which may be increased to as much as 17 g per square metre per annum within certain forest soils. Nitrogen inputs and outputs are very different. Much less than 30 % of the nitrogen added from above appears at the base whereas the inputs and outputs of sulphur often balance because many of the sparse soils in sensitive areas (e.g. Scandinavia) are fully saturated with sulphate but have very little nitrogen of their own.

Under natural conditions, sulphate is often a major anion in soilwater. Faster leaching of sulphate necessarily increases the loss of cations such as potassium, magnesium and calcium. Similarly, increased input of acidity increases the exchange between protons, on the one hand, and potassium, magnesium and calcium ions on the other; causing them to be more readily available later for leaching. Thus those soils most sensitive to acid rain will be non-calcareous, sandy soils naturally around pH 6 and low in colloidal material

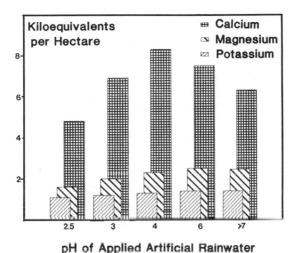

pH of Applied Artificial Rainwater

Fig. 4.3 Effect of simulated acid rain of different pH upon cation content of a Scandinavian forest soil (after Abrahamsen, 1980, see Beilke and Eshout, 1983).

through which water percolates easily. Nevertheless, acid soils with low cation content may show the largest proportional loss of nutrient ions even though the pH change is minimal. Figure 4.3 shows the effect of increasing acidity on the amounts of extractable potassium, magnesium and calcium remaining in iron podsol soil beneath pine forests which illustrates this point.

Within most soils there are various buffering systems which resist changes in acidity. In natural soils, classic buffering is carried out by a variety of different buffer systems, some of which are listed in Table 4.3. One of these, involving the dissolution of calcium carbonate (Reaction 4.17), has essentially been described before (Sect. 4.1b) but carbonic acid–bicarbonate exchange also takes place with various silicates of which Reaction 4.18 is just one example.

Table 4.3 Typical buffer systems operative in most soils

pH buffer range	Buffer system
8–6.2	Calcium carbonate–bicarbonate
6.2–5	Silicate–bicarbonate exchange
5–4.2	Cation exchange systems (e.g. Ca^{2+}, Mg^{2+}, NH_4^+)
4.5–2.8	Hydrated aluminium hydroxide exchange
3.8–2.4	Iron exchange

$$CaCO_3 + H_2O + CO_2 \text{ gas} \rightleftharpoons Ca^{2+} + 2HCO_3^- \qquad (4.17)$$

$$CaAl_2Si_2O_8 + 2H_2CO_3 + H_2O \rightleftharpoons Ca^{2+} + 2HCO_3^- + \qquad Al_2Si_2O_5(OH)_4 \qquad (4.18)$$

In more acidic soils of pH range 2.8 to 5 there are a series of cation-exchange–neutralisation reactions which ultimately involve the dissolution and precipitation of aluminium. Natural forms of aluminium in rocks and soils are highly variable, existing as silicates, alums, oxides and a wide variety of other forms. Weathering processes accelerated by acidic rain, cause exchangeable aluminium hydroxide to become more readily available upon the surfaces of soil and rock particles. Normally, exchangeable basic ions like calcium and magnesium are rapidly exchanged from rock and soil by protons in a reversible and rapid reaction which serves to buffer soil pH (Reaction 4.19). If this process experiences excessive inputs of acidity then it may become limited by the availability of exchangeable basic divalent atoms like calcium, magnesium, etc. Once this occurs, then release of trivalent aluminium may also serve to consume protons.

$$(\text{soil})-Ca + 2H^+ \rightleftharpoons (\text{soil})-2H + Ca^{2+} \qquad (4.19)$$

$$Al^{3+} + OH^- \rightleftharpoons Al(OH)_3 \rightleftharpoons M^+ + H_2AlO_3^- \qquad (4.20)$$

As Reaction 4.20 shows, aluminium hydroxide is capable of having both basic and acidic properties in the presence of strong bases or strong acids and is thus called an 'amphoteric electrolyte' (or ampholyte). The mobilisation of aluminium hydroxide from soil surfaces is therefore more accurately described as a 'dissolution' (Reaction 4.21) followed by a proton transfer (Reaction 4.22). Hydrated aluminium ions are really cationic acids and they are transferred not to a very weak acid (H_2O) but to an acid of moderate strength. Further transformations (Reactions 4.23 and 4.24), and if necessary the appropriate dissolutions, may follow if required. Consequently acidity does not disappear but is transferred to a progressively weaker cationic acid.

$$Al(OH)_3 + 3H_2O \rightleftharpoons Al(OH)_3(H_2O)_3^0 \qquad (4.21)$$

$$Al(OH)_3(H_2O)_3^0 + H^+ \rightleftharpoons Al(OH)_2(H_2O)_4^+ \qquad (4.22)$$

$$Al(OH_2)(H_2O)_4^+ + H^+ \rightleftharpoons Al(OH)(H_2O)_5^{2+} \qquad (4.23)$$

$$Al(OH)(H_2O)_5^{2+} + H^+ \rightleftharpoons Al(H_2O)_6^{3+} \qquad (4.24)$$

If excess acidity still remains then limited mobilisation of iron oxides may take place as well as loss of aluminium from the soil as the hydrated forms of Al^{3+}, $Al(OH)^{2+}$ and $Al(OH)_2^+$. These cations then appear in the run-off waters draining into streams, rivers and lakes.

Such a process occurs especially quickly upon very hard granitic regions (as in southern Norway) which experience large falls of acidic rain but where weathering rates are very low. This means there is little exchangeable basic cationic material like calcium to counteract the acidity.

The soil–water system is thus extremely complex but the processes that affect acidity are essentially threefold: exchange, buffering and neutralisation. During exchange, increased levels of hydrogen ions cause either a displacement and release of adsorbed basic cations (calcium, magnesium, potassium, sodium or ammonium) or they take up new exchange sites. If there is a strong counterbalancing anion in excess, like sulphate, then basic cations may be transported outward and lost. Buffering takes place by virtue of exchange reactions involving bicarbonate and aluminium ions that have already been described (Reaction 4.17 to 4.24) and neutralisation involves the formation of salts from acidic inputs reacting with bases within the soil. Many factors are involved in the various neutralisations that are possible but both soil composition and temperature are the most important variables. Inevitably, the amount of available bases within certain soils is correspondingly reduced but, in certain circumstances, acidification may cause an increase in the rate of mineral weathering and thus a fresh release of calcium and magnesium. Alternatively, the reaction products of weathering and neutralisation may build up in close proximity to surfaces or exchange sites to prevent any further increase in the rate of weathering and even decrease these rates after time. Clearly, with so many soil types and variability of conditions from place to place, it is difficult to predict and, more importantly, extrapolate these micro-effects within soils on to a wider global scale. As Ch. 7 goes on to show, at least one different theory to explain 'forest decline' (or 'dieback') is based primarily upon fundamental changes in forest soils such as those described here which may then cause detrimental effects upon trees through their root systems.

Acidic groundwater is known to cause considerable dissolution of buried metal pipework and it has also been suggested that heavy metals may be mobilised directly from soils by virtue of increased acidification. Most of these heavy metals are normally not very mobile although their solubility does increase with increasing acidity. They normally tend to bind strongly to organic material and only show mobility when humus levels are low and levels of acidity particularly high. Lead ions are more widely distributed because lead tetra-ethyl is added to many road traffic fuels to act as an anti-knocking agent and therefore lead is emitted in exhaust gases. Indeed, lead has also been suggested as an additional cause of forest decline (see Ch. 7). Table 4.4 shows how different heavy metals changed over a year in a spruce forest soil in West Germany. As may be observed, all were

Table 4.4 Annual heavy metal deposition into and output by seepage from a spruce forest ecosystem at Solling, West Germany (after Meyer, see Beilke and Elshout, 1983)

Element	Total deposition	Soil seepage output
	(kg per ha per annum)	
Aluminium	2.8	24
Cadmium	0.02	0.03
Copper	0.66	0.11
Chromium	0.17	0.006
Iron	2.1	0.16
Lead	0.73	0.013
Nickel	0.14	0.07
Zinc	1.7	2.4

accumulated with the exception of aluminium and, interestingly, zinc. Lead, on the other hand, showed little inclination to emerge in run-off waters.

Finally, mention must be made of the ammonium ions which reach the soil when dissolved in raindrops or whilst contained in aerosols in the atmosphere containing ammonium sulphate, ammonium nitrate and ammonium carbonate which come in contact with the ground. This acquisition is directly opposed by the process of ammonia volatilisation (see Sect. 3.1f) whereby ammonia escapes from various land surfaces into the atmosphere. The equation governing the equilibrium between ammonia and ammonium (Reaction 4.25) is heavily in favour of the formation of ammonium and against the release of ammonia at soil pH values lower than 8 even though the pKa of Reaction 4.25 varies with temperature, humidity and the carbon dioxide concentration. According to Reaction 4.25, any ammonia volatilisation that does occur must cause further acidification in the soil from which this occurs. Fortunately, this acidification is only a problem of alkaline soils with high calcium content (i.e. lime-treated arable land, see sect. 3.1f). The major effect of additional inputs of ammonium from rainfall and aerosols to acidic soils is to provide an additional basic exchangeable cation (i.e. ammonium) to assist with the neutralising capabilities of calcium and magnesium.

$$NH_3 + H^+ \rightleftharpoons NH_4^+ \tag{4.25}$$

There is one school of thought that holds that changing agricultural and forestry practices in remote regions have contributed to the acidification of groundwaters. In the past, upland pastures were often

heavily manured in spring and the ammonium added helped to
neutralise the groundwaters. With increasing rural depopulation many
of these pastures have now been abandoned and the practice is much
reduced. Furthermore, changes in forestry practice have taken place
so that intensive planting and thinning have increased the amount of
needle litter falling on to the land. These are potent sources of acidity
in their own right which could increase groundwater acidity still
further. How much these two changes in land use have contributed
to problems associated with acidic deposition is not known. Long-
term experiments are now in progress to investigate their relative
contributions to the overall mechanism of acidification.

(b)   Evaporation and sublimation

There are a number of mechanisms by which precipitated acidity
actually increases hydrogen ion content, and therefore, causes damage
or deterioration before rainwater enters soil–water systems. Figure
4.4 is an attempt to illustrate these processes. If rainwater falls on

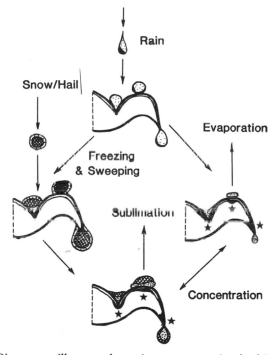

Fig. 4.4   Diagram to illustrate the various processes involved in the
concentration of acidity upon leaf and rock surfaces. The asterisks show
where greatest corrosion is most likely to take place.

exposed vegetation, rock or other material, it may flow easily away, enter a fissure or hole or remain held as a water droplet by virtue of surface tension. The latter is even more likely if the receiving surface is hydrophobic (i.e. water-repelling). Waxy leaf cuticles are good examples of such surfaces although there are also the possibilities of pits (e.g. sunken stomata) or hairs, etc., to catch and retain droplets especially if the droplet sizes are small as in mists. Such droplets or fluid-filled crevices may be subjected to various microclimatic conditions. For example, the water in them may freeze or it may be slowly evaporated by the sun or wind once the precipitation ceases. This has a most marked effect on the ionic composition of such droplets. As water molecules are lost, the concentration of ions (including protons) rises and would cause the local pH to fall unless prevented by adequate buffering. If not, highly localised concentrations of acidity could be available to attack surrounding structures such as the waxy cuticles of leaves, or fabrics and materials useful to man, or exposed rock surfaces. These microlocations, in turn, provide focal points for a repetition of this sequence of events if precipitation reoccurs, with the effect that damage is focused upon critical points.

If the precipitated droplets freeze, then natural weathering of exposed rocks takes place. But at least two other acid-concentrating processes may also occur. Firstly, the process of ice formation (involving only water molecules) has the effect of sweeping any other contents of the droplet ahead of the forming ice crystals. This means that elevated levels of other chemical species, e.g. ions, etc., are to be found in restricted parts of the droplet as the whole structure stabilises. If the temperature fluctuates around 0 °C and the process is repeated many times, this sweeping effect may be quite marked.

However, if the air above the frozen droplet or fallen snow particles is warmer or drier, then the process of sublimation takes place. This occurs when water molecules in the solid phase move directly into the gaseous phase leaving behind a greater concentration of contaminating ions, etc. (including protons) in a similar manner to the process of concentration by repeated evaporation mentioned earlier. The consequences of increased local acidity by such processes may be initially less than that of unfrozen acidic water droplets themselves because the lower temperature may reduce the rate of attack on the surface beneath. Once melted, however, the effect will be sudden and could even be more damaging because the concentration may have been allowed time to build up to even higher levels of acidity (lower pH) by repeated ice crystal sweeping.

It has been suggested that water droplets do not even need to land on a surface to undergo this progressive acidification. Mists may become acidified, especially at high altitude, by a combination of enhanced oxidation and evaporation or sublimation processes taking place whilst the droplets are in suspension. Subsequently, they adhere to vegetation when they come in contact by the process known as 'occult deposition' or 'droplet interception'.

All these phenomena will have their greatest effect at the tips of conifer needles or on the upper branches or crowns of trees. Such simple physical processes as these are obviously alternative candidates for a partial explanation of certain features of 'forest decline' (see Ch. 7). Some very low pH determinations (less than pH 1) have been recorded in ice–water droplets on conifer needles in central Europe where repeated evaporation–sublimation cycles are known to have taken place either on the needles or in the mists that swirl around them.

(c)   Acidity in rivers and lakes

Bicarbonate is the most important component in terms of the capacity of large bodies of freshwater to neutralise the effect of hydrogen ions. High concentrations of bicarbonate occur in so-called 'hard' waters, but in waters that fall below pH 5.4 bicarbonate is completely absent and the description, 'permanently acid' is used to discriminate such waters from soft or very soft waters which do contain small amounts of bicarbonate (Fig. 4.5). There is also a biological distinction between permanently acid and very soft natural waters because the pH zone between 5.7 and 5.2 marks a distinctive boundary between the survival and elimination of many freshwater animals.

Many studies have now been carried out which illustrate the increasing proportion of permanently acid lakes in sensitive areas in Scandinavia or North America. Table 4.5 illustrates the typical trend towards acidification in southern Norway. In practice it is always difficult to calculate rates of acidification of lakes because climatic, geographical and geological facts are so variable. The best means of expressing global acidification trends is to use analyses of Arctic ice which show that over the last 30 years there has been a regular decrease of 0.007 pH unit per annum which is directly attributable to man-induced emissions of pollution. Measurements of polar ice also show that greatest acidification occurs during spring which coincides with the increased movement of polluted air into the polar regions at the end of the long polar winter. In summer, the Arctic is again protected by the movement of the polar frontal weather system well to the north of the industrialised regions. As yet, no definite trend has been observed in Antarctic snow measurements over the last 25 years.

Measurement of pH alone is not an accurate description of the

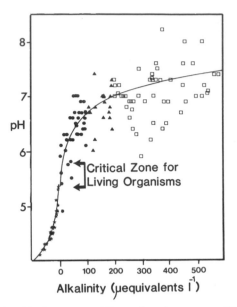

Fig. 4.5 Relationship between acidity and various types of hardness to be found in the upland lakes of the Lake District, UK in 1983: asterisks, permanently acid; solid circles, very soft; solid triangles, soft; open squares, medium hard waters. (Courtesy of Drs Sutcliffe and Carrick and the Freshwater Biological Association.)

Table 4.5 Comparison of pH in eighty-seven lakes in southern Norway with time (after Wright, 1977)[†]

pH range	Number of lakes[‡]	
	1923–1949	1970–1976
4.5 and below	0	2
4.6–5	3	25
5.1–5.5	18	14
5.6–6	21	17
6.1–6.5	11	15
6.6–7	20	13
7.1–7.5	17	9
7.6–8	3	5
8 and above	7	0

[†] SNSF Report TN 34/77, p. 71.
[‡] Expressed as a percentage.

acidic properties of natural waters. The hydrogen ion concentration of acidified lakes depends not just on the concentration of carbonic acid but also on the concentrations of strong bases, of other weak organic acids and, above all, on the concentration of the deposited strong acids. To measure all of these independently and accurately is technically difficult and requires correction for volume changes during titrations. At the same time, the influence of the atmosphere above has to be taken into account. If one regards the hydrogen ion concentration as the non-neutralised residue from sulphuric and nitric acids then gaseous concentrations of sulphur dioxide and nitrogen dioxide above as well as relative humidity must be known. The major neutralising agent which enters lakewater directly from the atmosphere is ammonium. Consequently, knowledge of the gaseous concentration of ammonia as well as all the aqueous components, is relevant if it is important to establish how far a particular atmosphere–lakewater system is from equilibrium. Not surprisingly if all these are to be taken into account, large equations and extensive calculations by computers are required. From studies so far it has been shown that at least two organic acids with pK_a's of 4.65 and 3.55, as yet not positively identified, have been demonstrated to be present which contribute to the acidity of many lakewater systems.

4.3   PLANTS

(a)   Leaf injury and cuticular weathering

Occurrence of injury to foliage as a direct consequence of acid rain (or acidic wet deposition, precipitation, etc.) remains controversial. Damage to leaves of sensitive crop plants like radishes, beets, soyabeans and kidney beans has been observed when leaves have been exposed to simulated rain of pH 3.4 or lower, unlikely though this may be under natural circumstances.

The degree of injury by acid rain depends to a large extent on the effective dosage which is a function of both the concentration and the period of exposure. Thus the contact time of acidic droplets of films on the leaf surface determines the extent of the damage. Factors such as temperature, humidity, wind turbulence, surface wettability and leaf morphology will all influence the period of contact. It has been observed on sensitive plants, for example, that 95 % of all foliar lesions due to acidic precipitation occur where water normally accumulates on leaves: along the side of leaf veins or margins, near hairs and close to stomata.

Within these wetted areas (droplets or films) a number of processes may take place. The oxidation of dissolved sulphur dioxide may still continue (see Sect. 4.1c) and be accelerated by the concentrating processes involving evaporation–sublimation mentioned in Sect. 4.2(b). Such processes will also increase local acidity which then causes weathering or degradation of the waxy leaf cuticle. Strong acids are able to oxidise and hydrolyse the wax esters and release of some of the long fatty acid chains from the waxy matrix. This changes the water-repelling (hydrophobic) characteristics of the leaf cuticles and increases their wettability.

Leaf injury in the field, similar to that induced by artificial acidic treatments under controlled conditions, has not been positively detected and therefore cannot be unambiguously attributed to acid rain. Yet reports of abraded waxy cuticles and changes in the microscopic organisation of leaf surfaces have often been detected, especially on conifer needles after some considerable time even though no specific changes due to one specific acid precipitation incident have been positively identified. Indeed, field-grown plants appear to be much less susceptible to the development of foliar symptoms similar to those occasionally shown by experimental plants treated with simulated acid rain. However, the acid-induced long-term deterioration of the cuticular barriers of conifer needles, or cracking of the thin waxy plugs which cover the stomata of conifers, may allow direct penetration of acidic and other pollutants, enhanced insect and fungal attack, increased susceptibility to water loss, frost damage or other stresses. Nevertheless, on balance, it would appear that the direct effects of acidic wet deposition on leaves and needles are not a primary explanation of 'forest dieback' or 'decline' (see Ch. 7) although it may be an additional stress factor and may be capable of causing some long-term reductions in growth of both trees and crops.

(b) Physiological and biochemical changes

Studies of physiological and biochemical change due to acidic wet deposition, as distinct from that caused directly by dry deposition of sulphur dioxide (Sects. 2.2b and c) or oxides of nitrogen (Sect. 3.2c), have been few and as yet, by comparison, primitive. Quite a number have reported changes (mainly reductions) in pigment levels but little else. Indeed it has been very difficult to demonstrate changes in rates of photosynthesis or respiration without having to use very low and unrealistic levels of pH (e.g. pH 2). This makes it very hard to explain the claimed reductions in growth of sensitive crops like radishes, beets and certain varieties of maize and soyabeans which have been attributed to acidic wet deposition of less than pH 4.

One possible physiological change which has often been suggested could operate by means of alterations in intracellular pH levels. The importance to plants of adequate *in vivo* pH buffering or regulation by 'pH-stat' mechanisms has already been emphasised in Ch. 2 (Sect. 2.2b). In studies at the University of Lancaster, we have shown that plants (barley and conifers) have the capability of resisting cellular pH change even when exposed regularly to simulated rain down to pH 3. Plant cells, unlike animal cells, have internal large fluid filled non-living spaces called 'vacuoles' which have a variety of functions including storage of wastes. Thus the vacuolar pH, which might be expected to change most if 'pH-stat' mechanisms were operating, changes by about a third of a pH unit in response to simulated acid rain although the buffering capacity of the leaves or shoots is unaffected. Research elsewhere has shown that size and nature of buffering capacity is an inherited characteristic. Consequently, acid-tolerant plants may be better at resisting pH change and, conversely, acid sensitive plants are less able to adapt. It is possible that the 'energy-redirection' hypothesis (see Sect. 1.5) might be one means of explaining net growth reductions of crop plants due to acidic wet deposition. As more pressure is put upon cellular systems to resist pH change, more energy is devoted to make the 'pH-stat' proton pumps work harder to push protons outwards from cells (or across the tonoplast membrane inwards into the vacuoles) as well as in storing more exchanged or buffered products. This may mean that there would be less energy for growth.

(c) Foliar leaching and reproduction

Applications of simulated acidic rain on trees such as maples, pines and spruce or on crops such as beans have shown that the leaching of cations like potassium, magnesium and calcium from leaves has been increased but, as Sect. 4.2(a) has shown, these may accumulate at the bases of plants along with other nutrients. In turn, the enhanced uptake of such nutrients by root systems may offset such losses caused by leaching.

Seed germination, early seedling growth, flowering and seed formation have been predicted, and in some cases demonstrated, to be sensitive to enhanced acidification although little is known about effects on flower or cone development. The full significance of this for forest regeneration and replacement of species has yet to be determined. Undoubtedly, any individuals within a population showing sensitivity to acidity at any stage of development would be preferentially eliminated. With such long life cycles in forest trees, it is very difficult to predict the long-term consequences like these to such ecosystems.

4.4 ANIMALS

(a) Effects on water bodies and fisheries

As shown in earlier sections, after precipitation has fallen it may be concentrated by evaporation or sublimation, be modified as it flows over vegetation, down tree trunks, etc., and altered as it percolates through soil. Decay of vegetation, principally leaves and needles, will contribute acidity as microbial activity breaks down organic debris both on the surface and deeper down in the soil. Subsequent exchange of cations and anions between soil particles may be extensive and alterations in the water table cause water-logging which then promotes microbial anaerobic reduction (see Ch. 2). Alternatively, drying out could cause a re-oxidation of sulphide to sulphate. Reductive processes like those converting nitrate to ammonia reduce acidity just as nitrification (see Ch. 3) or the oxidation of sulphide to sulphate promote its formation.

Over a short period (less than 10 years) in most affected areas, there are only small differences measured between inputs of sulphur compounds into catchment area and the amount of sulphur appearing in run-off waters (i.e. the whole system is saturated). Most studies have usually shown differences of less than 10 %. Barren bedrock, for example, with little soil shows small capacity for retention of sulphur compounds but peaty soils may still show discrepancies between input and output of sulphur of more than 40 %.

With run-off waters there may also be wide variations in acidity. Generally, as streams flow through limestone or chalk catchment areas, they become less acid, whilst lakes show pH variations with depth or wide pH differences between inlet and outlet streams. Seasonal variations may be pronounced. The period of snow melt is often associated with high levels of acidity. Figure 4.6 shows a fairly typical experience in Scandanavian river systems characterised by a fall in pH immediately after the melting of snow which is then followed by an increase in the levels of aluminium ions released from soil–rock systems as a consequence of the acidity (see Sect. 4.2a).

Short-term effects of acidity upon fish are listed in Table 4.6 but these are only intended as a rough guide as there is considerable variation between species or individuals and at different stages in their life cycles. Other factors besides acidity are important. Levels of carbon dioxide, calcium chloride, iron and trace metals all cause variation in response. The most critical ions, other than protons, are undoubtedly those of aluminium (see Sect. 4.2a). Many studies have shown that aluminium ions are toxic to fish and that the extent of this toxicity is related to the level of acidity. Studies on trout, for example, have shown that levels of $90 \, \mu\mathrm{g} \, \mathrm{l}^{-1}$ aluminium (the highest level

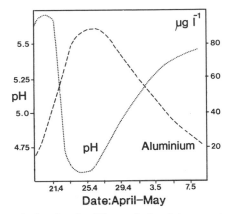

Fig. 4.6 Changes in levels of acidity and aluminium content of a Scandinavian river system which illustrate a typical spring flush event

recorded during the spring flush event shown in Fig. 4.6) at pH 5.2 allow trout to live longer than 14 days. However, an increase of acidity (to pH 4) but with similar levels of aluminium causes most fish to die after 5 days even though no gill damage is observed.

Calcium levels, in the presence of aluminium ions, are critical. If calcium concentrations are low then aluminium is toxic, but when

Table 4.6 Short-term effects of acidity upon fish (after Alabaster and Lloyd, 1980)[†]

pH range	Effect
6.5–9	No effect
6.0–6.4	Unlikely to be harmful except when carbon dioxide levels are very high ($>1\,000\ mg\ l^{-1}$)
5.0–5.9	Not especially harmful except when carbon dioxide levels are high ($>20\ mg\ l^{-1}$) or ferric ions are present
4.5–4.9	Harmful to the eggs of salmon and trout species (salmonids) and to adult fish when levels of Ca^{2+}, Na^+ and Cl^- are low
4.0–4.4	Harmful to adult fish of many types which have not been progressively acclimated to low pH
3.5–3.9	Lethal to salmonids, although acclimated roach can survive for longer
3.0–3.4	Most fish are killed within hours at these levels

[†] *Water Quality Criteria for Freshwater Fish*, pp. 21–45. FAO, Butterworths, 1980, London.

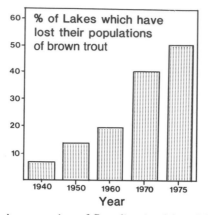

Fig. 4.7 Increase in proportion of Scandinavian lakes (2850 in total) which have lost their populations of brown trout.

calcium is high then even at low pH, aluminium has no effect. The problem is all the more complicated because newly hatched fry are most sensitive to acidity whereas the later 'swim-up' fry are highly sensitive to aluminium. Thus careful experimental attention to the interactions between life stage, and levels of acidity, aluminium and calcium is essential for a full understanding of these problems. This is perhaps the most critical area of uncertainty that remains, if a linkage is to be established conclusively between the effects of likely changes caused by acid rain and reductions in freshwater populations of fish so widely reported over the last two decades. In 1983, for example, it was estimated that lakes in southern Norway representing over 13 000 km^2 of surface area were practically devoid of fish. Figure 4.7, which shows the proportion of those which have lost their brown trout over a period of 35 years, is taken from a survey of 2850 lakes in southern Norway and illustrates the extent of the problem.

One additional problem has also arisen recently which could have long-term problems even if liming of affected waters is undertaken at regular intervals. Phosphate is a critical nutrient for fish growth but levels of phosphate in lakewater and riverwater in acid-affected regions remain persistently low. It would appear that, in catchment soils, aluminium precipitates the available phosphate close to run-off systems as insoluble aluminium phosphate. Liming programmes directed at lakewaters will therefore have to take this phenomenon into account and include appropriate amounts of extra phosphate.

(b) Mechanisms of toxicity in fish

Blood plasma of freshwater fish contains high levels of both sodium

Water Uptake &
Active Uptake of
Salt through Gills

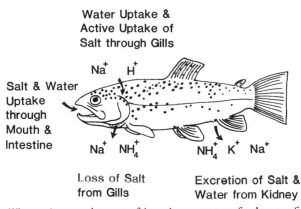

Salt & Water
Uptake
through
Mouth &
Intestine

Na^+ H^+

Na^+ NH_4^+

NH_4^+ K^+ Na^+

Loss of Salt
from Gills

Excretion of Salt &
Water from Kidney

Fig. 4.8 The various exchanges of ions important to freshwater fish.

and chloride (equivalent to 150 mM l^{-1} NaCl) and lesser amounts of calcium, potassium and bicarbonate. Soft freshwaters, which are more prone to acidification, contain less than 0.1 mM sodium chloride yet fish must replace any losses in urine or from the gills by uptake mainly through the gills. The major exchanges are indicated in Fig. 4.8.

The architecture of fish gills is complex. Gill slits divide the sides of the throat into distinct gill arches. From these project gill filaments which have individual gill lamellae covered by different cells, most of which are respiratory surface (epithelial) cells and mucous cells. The filaments are covered with mitochondria-rich cells which are analogous to the 'chloride' cells of seawater fish although they do not have a similar function. Respiratory lamellae are supplied by arterial blood whilst mitochondria-rich cells are served by the venous system. The former areas are largely responsible for uptake of sodium and chloride whilst the latter recover calcium. Uptakes of both sodium and chloride in fish gills are both energy-requiring (i.e. active) processes which are quite distinct from each other. Both types of cross-membrane pump appear to use ATP as an energy source although knowledge concerning the chloride mechanism is uncertain. At the moment it is thought that both hydroxyl and bicarbonate ions are counterexchanged outwards at the same time as chloride is drawn in.

In neutral waters, the sodium pumps are able to capture sodium against a thousandfold difference when levels of protons are the same both inside and out. In acid waters (e.g. pH 4), however, counterbalancing hydrogen ions must be expelled against a thousandfold gradient and the work load is obviously much greater. Because such large concentration gradients exist in both directions, advantage is always taken of ammonium ions inside and these are ejected wherever possible in place of protons.

125

Gill membranes are naturally partially permeable to sodium – hence the necessity to replace the high body levels. They are also highly permeable to hydrogen ions. Even if the latter are ejected by active pumping they may also return passively (i.e. without hindrance and energy expenditure). However, calcium ions have the ability to reduce this easy access. In other words, at high calcium levels, less protons are able to enter and less sodium ions are able to leave. Furthermore, bicarbonate, one of the counter-ions for chloride uptake, is converted to carbon dioxide if acidity rises but this also makes the active entry of chloride more difficult. Calcium, however, also reduces the permeability of gill membranes towards chloride ions – so less are lost and more protection against acidity is afforded.

The primary cause of fish death in acid waters (Table 4.6) is excessive loss of critical ions such as sodium which are not replenished by pumping. When blood plasma sodium and chloride levels fall by about a third (to the equivalent of 100 mM NaCl), body cells swell and extracellular fluids become more concentrated. Losses of potassium from cells may partly compensate for these changes but if the potassium is not eliminated from the body fast enough then depolarisation of nerve and muscle cells takes place. Uncontrolled twitching of non-acclimated fish suddenly exposed to acid waters before death is symptomatic of these effects.

It is therefore calcium which appears to assist fish to cope with acid waters and to allow them to acclimatise to acidic conditions. These ions naturally bind to external surfaces of gills. After adaptation by fish to acidic waters, the affinity of calcium for the membrane surfaces of their gills increases. Sulphate appears to be much more effective than nitrate or chloride in displacing calcium from these binding sites at pHs above 3.8 but below this there is very little difference. Indeed, calcium is now recognised as one of the major regulator molecules of all living systems even at very low levels in the cell. The modulation or regulation of membrane permeability mentioned above is just one of the cellular events controlled by these ions.

The additional presence of aluminium ions appears to interfere with the regulation by calcium of permeability and other osmotic processes. A number of studies have shown that aluminium is toxic to fish at certain pH levels around 5 to 5.5 but less so at higher or lower levels of acidity. Undoubtedly, conditions of spring flush (Fig. 4.6) bring about levels of pH around 5 and higher concentrations of aluminium ions (mainly $Al(OH)^{2+}$) which are so harmful. Aluminium appears to act in a number of ways. It promotes efflux of sodium ions at those critical pH levels when the proportion of monovalent $Al(OH)_2^+$ (more accurately $Al(OH)_2(H_2O)_4^+$) is highest. If the pH is lower then more $Al(OH)^{2+}$ and Al^{3+} ions are present and the losses of sodium are less. Recent work would suggest that aluminium in the form of $Al(OH)_2^+$ also causes mucous clogging of gills, as well as interfering with respiration, and with other regulatory events associated with calcium. Such ions are now known also to inhibit developmental events which prevent the essential calcification of the skeleton if fish fry are to mature properly. It is now thought that losses of immature fish fry due to these causes and the failure to grow through to maturity are some of the prime causes of the long-term loss of fish stocks.

(c) Effects on invertebrates and birds

Most species of mayflies, caddisflies, freshwater shrimps, limpets, snails and beetle larvae are absent from acidic water systems when the pH falls below 5.4. Early suggestions that decreases in numbers due to progressive increases in acidity cause the fish that depend upon certain invertebrates to starve are now generally not thought to be major explanations for the decline of fisheries. The decline in invertebrate populations occurs in parallel with that of fisheries and, it is thought, through very similar disturbances in pH control, osmotic regulation or calcium and aluminium interactions already shown in fish. More fundamental physiological studies on invertebrates are urgently required in this area to confirm this.

Aluminium and heavy metals released by acidification of run-off waters are concentrated by many invertebrates and thus enter food chains. Many birds which are reliant upon these creatures therefore receive secondary doses of aluminium and heavy metals in their diets. Their numbers are reduced in such areas mainly because of imperfect calcification of their eggs which fail to produce live fledgelings.

(d) Direct or indirect effects upon humans

There are no known direct effects of acid rain on humans but they do occasionally experience sudden industrial exposures to aerosols or mists containing commonly used acids such as sulphuric, nitric, hydrochloric and hydrofluoric acids. The last named are covered separately in Ch. 6 but accidental exposure to acidic aerosols is a well recognised hazard. The symptoms are, by and large, similar to those of their parent gases, e.g. sulphur dioxide (Ch. 2) and nitrogen dioxide (Ch. 3), each of which induces various irritations and pulmonary disturbances. The main differences concern the size of the aerosol droplets involved in any incident. It would appear that droplet sizes of 0.8 μm or less are the most harmful. In other words more lung damage may be caused by a lesser amount of acidity if it is more finely divided.

Indirect effects of acid rain could be more harmful to the population over a longer time period through changes in water quality. By far the largest contamination of natural waters by nitrate arises from excess run-off from fields which have been treated with artificial fertilisers. It is therefore unlikely that acidic precipitation upon water catchment systems has any further significant impact upon these inputs of nitrate. The suggestion of a linkage between high-nitrate water supplies and the occurrence of stomach cancer is contentious. Certain mouth bacteria can turn nitrates into carcinogenic nitroso derivatives and individuals with unusually high-nitrate diets do tend to have higher rates of stomach cancer. Even though this linkage between high-nitrate water supplies and stomach cancer has not been firmly established, the European Community has taken the precaution of limiting nitrate levels in public water supplies to a maximum of 50 mg l^{-1}. Levels in drinking water in the Netherlands and the UK are already approaching, or in certain places exceeding, that level and methods are being evaluated to treat and remove nitrate effectively from such supplies. The most favoured remedy involves a combination of ion exchange to replace nitrate with bicarbonate and to use microbial denitrification to reduce the exchanged nitrate to nitrogen.

The most likely hazard from alterations to public water supplies by acidic precipitation is a general increase in the level of aluminium leached from soils in certain water-catchment areas. There has been an increase in the number of cases of a rare bone-wasting disease called 'osteomalacia' in areas where aluminium levels in water supples are high (1000–2000 ppm). A possible link between the two was drawn when patients suffering from kidney failure were once treated with aluminium which had the side-effect of bone softening. This, in turn, could be alleviated by the use of aluminium-chelating drug, desferrioxamine, which is also effective in treating osteomalacia. The possible interference of aluminium with calcification of bones in humans may be analogous to the improper calcification observed when fish fry mature in acidified, aluminium-enriched lakes and rivers.

Another disease attributed to the availability of aluminium in natural waters is Alzheimer's disease, a senile dementia that occurs much earlier in life than is normally the case and the sufferer dies about 10 years after the onset of the condition. Post-mortem examination has revealed deposits (plaques) containing aluminium within the brain. The interesting feature is that the aluminium is at the centre of focus of the plaques which implies aluminium is the cause of the plaque formation which then impedes nerve–nerve communication, hence the dementia. Other surveys have shown the possibility of a link between high incidences of various dementia conditions and low levels of calcium either in the diet or drinking water. In one study as many as 30 % of patients admitted to hospital with broken bones caused by calcium deficiency were also found to be suffering from dementia. This linkage may not yet be firmly established but removal of aluminium from water supplies, diets and cooking utensils is an obvious sensible precaution. In parts of Scandinavia it is now standard practice to remove aluminium from water used for kidney dialysis treatment and making up baby feeds. In some cases treatment of whole drinking-water supplies is now being introduced.

SELECTED BIBLIOGRAPHY

Beilke, S. and Elshout, A. J. (eds.), *Acid Deposition.* D. Reidel Publ. Co., 1983, Dordrecht.

Drabloes, D. and Tollan A. (eds.), *International Conference on the Ecological Impact of Acid Precipitation.* Sanderfjord, 1980, Oslo-Aas.

Duensing, E. E. and Duensing, L. B., *Environmental Effect of Acid Precipitation.* Vance Bibliographies, 1980, Monticello, Ill.

Howells, G., Acid waters – the effects of low pH and acid associated factors on fisheries. *Advances in Applied Biology* 9, 1983, 143–255.

Hutchinson, T. C. and Havas, M. (eds.), *Effects of Acid Precipitation on Terrestrial Ecosystems.* Plenum Press, 1980, New York.

Schneck, J. J., *Acid Rain: A Critical Perspective.* Tasa Publ. Co., 1981, Minneapolis, Minn.

Schrader, S., Greve, U. and Schönwald, H. R. (eds.), *Acid Precipitation and Forest Damage Bibliography.* Bundesforschungsantalt für Forst-und Holzwirtschaft, 1983, Hamburg.

The Watt Committee on Energy, *Acid Rain.* Report no. 14, 1984, London.

Toribara, T. Y., Miller, M. W. and Morrow, P. E. (eds.), *Polluted Rain.* Plenum Press, 1980, New York.

VDI-Kommission Reinhaltung der Luft, *Saure Niederschläge (Acid Precipitation).* VDI-Berichte no. 500. 1983, VDI-Verlag GmbH, Düsseldorf.

5

Ozone, PAN and photochemical smog

'And the winds and sunbeams with their convex gleams
   build up the blue dome of air'

from 'The Cloud' by Percy Bysshe Shelley (1792–1822)

5.1   FORMATION AND SOURCES

(a)   Monatomic oxygen, ozone and hydroxyl radicals

In the thermosphere 80 km and upwards above the surface of the
Earth, oxygen (see Fig. 1.4) exists almost exclusively in monoatomic
form because high-energy photons break up the most stable of
molecules (Reaction 5.1). At lower altitudes, oxygen atoms partially
combine to form diatomic (molecular) oxygen and triatomic ozone
(Reactions 5.2 and 5.3, where M represents an energy-absorbing
third body).

$$O_2 \ + \ \text{ultraviolet light} \ \longrightarrow \ 2O \tag{5.1}$$

$$2O \ + \ M \qquad \longrightarrow \ O_2 + M \tag{5.2}$$

$$O \ + O_2 + M \qquad \longrightarrow \ O_3 + M \tag{5.3}$$

Greatest concentrations of ozone are to be found in the stratosphere (15–40 km) where light causes some destruction (Reaction
5.4) of this ozone layer although this particular reaction is rather

Ozone removal by:

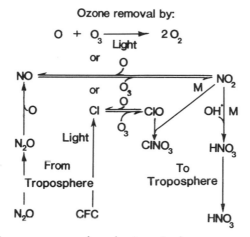

Fig. 5.1 Major ozone removal mechanisms in the upper atmosphere (stratosphere) include natural photolytic decomposition assisted by oxides of nitrogen formed either from high-flying aircraft or, more likely, by oxidation of nitrous oxide arriving from the lower atmosphere (troposphere). The latter arises mainly from microbial denitrification taking place in the soil which is enhanced by excess artificial fertiliser being applied to stagnant land. Another important mechanism of ozone removal is by means of chlorofluorocarbons, CFC (used as spray can expellants and refrigerants) being converted to oxides of chlorine and chlorine nitrate. It is important to appreciate that loss or uptake of monatomic oxygen is the equivalent to ozone removal because they are readily interconverted.

slow. However, the additional presence of oxides of nitrogen accelerates the disappearance of ozone (Reactions 5.5 and 5.6) and together these replace the dependence on Reaction 5.4 (Fig. 5.1).

$$O_3 + light \longrightarrow O + O_2 \tag{5.4}$$

$$NO + O_3 \longrightarrow NO_2 + O_2 \tag{5.5}$$

$$NO_2 + O \longrightarrow NO + O_2 \tag{5.6}$$

The most significant and reactive free-radical species of the atmosphere is the hydroxyl radical ($OH^{\bullet}$) formed by reaction of atomic oxygen with water (Reaction 5.7) in the troposphere lower down in the atmosphere. Hydroxyl radicals react readily with many of the other contaminating gases in the atmosphere, often in a long series or chain of reactions. The sequence of complete oxidation of methane to carbon dioxide involving three reactions with hydroxyl radicals is an illustration of this reactivity (Fig. 5.2). The last step of this sequence

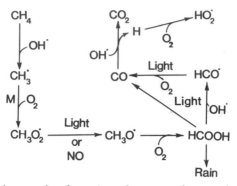

Fig. 5.2 A likely cascade of reactions that cause the complete atmospheric oxidation of methane to carbon dioxide to illustrate many of the probable reactions involving light, oxygen and hydroxyl radicals. The last oxidation of carbon monoxide to carbon dioxide is a very important reaction within the atmosphere.

is particularly rapid and the major mechanism by which carbon monoxide is removed from the atmosphere (see Sect. 6.3). Hydroxyl radicals and ozone also feature in reactions which are capable of removing oxides of nitrogen from the stratosphere (Fig. 5.1), some of which have already been covered in Ch. 3.

$$O + H_2O \longrightarrow 2OH^{\bullet} \tag{5.7}$$

In the mid-1970s, considerable concern was expressed over the possibility that large numbers of supersonic aircraft would inject significant quantities of nitric oxide and nitrogen dioxide into the ozone layer of the lower stratosphere. These pollutants could then cause enhanced destruction of ozone, thereby removing the protective shield that this layer afforded to the biosphere from incoming radiation, particularly in the ultraviolet region. For many reasons, technical as well as economic, the numbers of aircraft flights and projected supersonic developments did not occur at the time but subsequent computer simulations have shown that possible harmful effects from these sources might have been overestimated. Recent assessments suggest that the effect of direct injection of oxides of nitrogen from high-flying aircraft or from hypersonic developments is still likely to be small and stratospheric recovery relatively quick (i.e. three or so years).

The same cannot be said for two other potential threats to the ozone layers, nitrous oxide from artificial fertilisers (Ch. 3) and halocarbons or chlorofluorocarbons from aerosol cans, etc. (see below). The recovery time in each of these cases is from 20 to 50 years, too long for remedial action after the first intimation of any problem.

In the turbulent troposphere below the stratosphere the situation is rather different. The major source of ozone here is the photolysis of nitrogen dioxide (Reaction 5.8) producing monatomic oxygen which then forms ozone by virtue of Reaction 5.9. The additional presence of hydrocarbons, aldehydes and carbon monoxide may accelerate the initial photolysis (Reaction 5.8) by improving the rate of nitric oxide oxidation by means of peroxyl radicals (e.g. Reaction 5.10) and at the same time produce more reactive hydroxyl radicals. The various inter-relationships between the oxides of nitrogen and ozone are encompassed within Fig. 3.8.

$$NO_2 + light \longrightarrow NO + O \qquad (5.8)$$

$$O + O_2 + M \longrightarrow O_3 + M \qquad (5.9)$$

$$HO_2^{\cdot} + NO \longrightarrow NO_2 + OH^{\cdot} \qquad (5.10)$$

(b) Sprays, refrigerants and solvents

The last two decades have seen a dramatic increase in the production of halocarbon compounds containing chlorine, fluorine or bromine constituents, sometimes also called fluorocarbons or chlorofluorocarbons or CFCs. They are used widely as aerosol propellants, refrigerants, foaming agents and solvents. Originally thought to be completely inert and harmless, compounds such as $CFCl_3$ and CF_2Cl_2 (and occasionally including CF_4, C_2F_6, CCl_4, $CClF_3$, $CHClF_2$, $CHCl_2F$, $C_3H_3Cl_3$, $C_2Cl_3F_3$, $C_2Cl_4F_4$, C_2ClF_5, $CBrF_3$ and $CBrClF_4$) now amount to just over 0.1 ppbv of the troposphere. An amount which may actually account for almost the whole of the world's production since they were first introduced.

Problems with halocarbons start in the stratosphere. By a complicated series of reactions (not shown) they are broken down by light or by reaction with monatomic oxygen to produce chlorine atoms and oxides of chlorine. These, in turn, may act rather like the oxides of nitrogen to remove ozone (and monatomic oxygen) from the stratosphere (Reactions 5.11 and 5.12) by mechanisms similar to those of the oxides of nitrogen (Reactions 5.5. and 5.6) as shown in Fig. 5.1.

$$O_3 + Cl \longrightarrow ClO + O_2 \qquad (5.11)$$

$$ClO + O \longrightarrow Cl + O_2 \qquad (5.12)$$

Until recently, the levels of halocarbons and their rates of transport into the stratosphere were thought to be capable of causing a 5 to 10 % reduction in ozone after 100 years at current production levels.

A decade ago this estimate was even higher (33 %) and current predictions are still in a state of flux with possible revision downwards again because other possibilities exist. One conversion (Reaction 5.13), which occurs quite readily, may remove oxides of both chlorine and nitrogen from the stratosphere. This would then reduce the ozone depletion rates caused by both of these oxides separately. Furthermore, chlorine-containing hydrocarbons (mainly CH_3Cl and CCl_4) are released from natural sources at rates comparable to those emissions of man-made halocarbons. Moreover, this natural release has been going on over a much longer time period. In turn they also form oxides of chloride which then reduce ozone-depletion rates.

$$ClO + NO_2 + M \longrightarrow ClNO_3 + M \tag{5.13}$$

As has already been mentioned, it did appear at one time that atmospheric ozone-depletion hazards arising from aerosol (and from Concorde aircraft, solvents, refrigerators, etc.) might have been over-estimated. However, more recent work has shown that potential problems arising from atmospheric levels of man-made halocarbons cannot be disregarded. They are, on a molecular basis, as much as 10 000 times more efficient than carbon dioxide in absorbing solar energy reflected back from the Earth which would normally be lost to space. This 'greenhouse effect' is discussed at length in Ch. 6 but a rise of a few ppbv of these compounds on top of the global warming caused by elevated levels of carbon dioxide could be severe and accelerate the problem.

In terms of the mechanisms of ozone formation and the contributing presence of the oxides of nitrogen, the contrast between the troposphere and the stratosphere is quite marked. Oxides of nitrogen near the Earth promote ozone formation in the presence of strong light and hydrocarbons whereas higher up in the stratosphere they consume ozone and its surrogate, monatomic oxygen. The major input into the stratosphere from the troposphere is therefore nitrous oxide (N_2O) released by excessive microbial denitrification within soils which is caused by extra nitrogen being added to stagnant and poorly aerated land (Ch. 3). These inputs and their subsequent oxidation to nitric oxide which then consumes ozone (Reaction 5.5) might then affect the stability of the upper atmosphere. This, in turn, might damage the protection that the stratospheric ozone layer affords the surface of the Earth and the biosphere from harmful solar radiation, especially ultraviolet light, which is capable of damaging genetic material and thereby increasing the rates of mutation and the incidence of skin cancers. Current intimations of a 2–3 % reduction in stratospheric ozone cover could translate into 800 000 avoidable cancer deaths and 40 million cases of skin cancer.

(c) Unburnt hydrocarbons and photochemical smog

Evaporation of solvents and fuels, as well as incomplete combustion of fossil fuels, causes a different wide range of hydrocarbons to be released into the atmosphere. Analysis of some samples of ambient air have shown the presence of over 600 different hydrocarbons. The most abundant hydrocarbon in the atmosphere is methane which is released naturally from decaying vegetation, as well as from industrial and domestic sources, in amounts ranging from 1 to 100 ppmv. As Fig. 5.2 shows, atmospheric mechanisms exist for the complete oxi- dation of methane to carbon dioxide as well as for the breakdown of other simple organics. Indeed, some hydrocarbons found in the atmosphere may act as plant growth regulators (ethylene) whilst others (e.g. benzene) are known to be toxic to humans by inducing cancer of the blood system following prolonged exposure.

By far the greatest global problem with unsaturated hydrocarbons is their ability to promote the formation of photochemical smog in the presence of oxides of nitrogen, strong sunlight and stable meteoro-logical conditions. The reaction chains involved in photochemical smog formation may be long and complicated because one reaction involving a free radical often generates another, which in turn reacts to generate a third, and so on. The starting points for these chain reactions are numerous because of the variety of hydrocarbons released to the atmosphere but aldehydes and ketones easily produce free radicals in strong light (Reaction 5.14). This leads to the formation of peroxyl radicals $(RO_2^{\bullet})$ such as in Reaction 5.15. Ozone may also attack unsaturated hydrocarbons to produce similar free radicals (Reaction 5.16) as well as aldehydes to permit Reaction 5.14. Many of these free radicals generated in strong sunlight are thought to be mainly responsible for the eye irritation experienced by indi-viduals exposed to photochemical smog.

$$RCHO + light \longrightarrow R^{\bullet} + H^{\bullet}CO \qquad (5.14)$$

$$R^{\bullet} + O_2 \longrightarrow RO_2^{\bullet} \qquad (5.15)$$

$$O_3 + RCH = CHR \longrightarrow RCHO + RO^{\bullet} + HC^{\bullet}O \qquad (5.16)$$

Unfortunately, there are only a few reactions that terminate a chain of reactions instead of extending the sequences still further. Rarely are they stopped by a double free-radical collision (e.g. Reaction 5.17). Far more prevalent are those reactions where peroxyl radicals react with oxides of nitrogen to form a family of different peroxyacyl nitrates sometimes collectively abbreviated to PANs. (Note: although their formation may be expressed generally by Reaction 5.18, if R represents a methyl group then individually the product is called 'peroxyacetyl nitrate'; if R is equivalent to CH_3CH_2, then it is 'peroxyproprionyl nitrate', and so on). Peroxyacyl nitrates and similar substances are in themselves very reactive towards sensitive surfaces such as eyes or delicate plant tissues. The larger the carbon fragment of the peroxyacyl nitrate the more toxic it is. Some of the eye irritants appear to last longer than those agents that cause plant damage and probably involve other minor atmospheric contaminants like formaldehyde (Fig. 5.1) and acrolein as well as the peroxyacyl nitrates. As a rough approximation, one molecule of peroxyacyl nitrate is produced for every fifty ozone molecules.

$$RO_2^{\bullet} + R'O^{\bullet} \longrightarrow ROR' + O_2 \qquad (5.17)$$

$$R.CO.O_2^{\bullet} + NO_2 + M \longrightarrow R.CO.O_2^{\bullet}NO_2 \qquad (5.18)$$

Photochemical smog episodes have a distinctive daily rhythm (Fig. 5.3). In the absence of wind around dawn when urban activity is low, levels of carbon monoxide, nitric oxide and hydrocarbons only start to increase as the morning traffic builds up. Amounts of nitrogen dioxide rise to a maximum about $1\frac{1}{2}$ hours after the peak levels of nitric oxide when the sun is reasonably high. The later appearance of more ozone coincides with the disappearance of most of the nitric oxide molecules and at the same time aldehyde levels rise markedly after the hydrocarbon levels have peaked. Concentrations of ozone are therefore usually highest shortly after noon just after the levels of aldehydes have peaked. Still air conditions persisting at noon are then

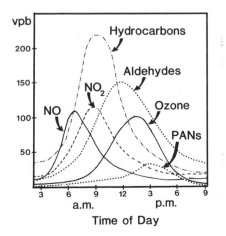

Fig. 5.3 A representative time-course of the development and decay of a photochemical smog over a large conurbation during a bright, warm working day with no wind.

ripe for free-radical chain reactions to form peroxyacyl nitrates and other irritants. Returning traffic in late afternoon generates more nitric oxide (and, thereby, nitrogen dioxide) which immediately scavenges the ozone and some of the aldehydes so levels of oxides of nitrogen do not rise to follow their distinctive biphasic daily pattern observed elsewhere (Fig. 7.1) especially under cloud cover. The blue–brown haze over heavily populated cities, often associated with photochemical smog, consists mainly of unburnt hydrocarbons in an advanced state of oxidation. Typically, measurements of total oxidising capability, arising from both the ozone and from the organic peroxides, are highest in the mid-afternoon and by midnight have almost disappeared.

Recent work has shown that such conditions necessary for the formation of this type of 'Los Angeles' photochemical smog actually exist throughout large parts of the Northern Hemisphere (and, in some instances, in the Southern Hemisphere as well). The main requirements are roughly equal atmospheric levels of oxides of nitrogen and hydrocarbons. These include acetylene, benzene, butanes, ethane, hexanes, pentanes, propane and toluene – all of which are characteristic of man-induced emissions which mainly occur north of 40°N. Under conditions of bright sunlight and still air, tropospheric levels of ozone (as well as PANs) produced from oxides of nitrogen and unburnt hydrocarbons are therefore more likely to arise from mobile rather than static sources. Legislation to control exhaust gases from vehicles and cut down emission of these primary contaminants will consequently reduce concentrations of this type of harmful photochemically generated pollution in areas susceptible to photochemical smog formation. Cost–benefit analyses have shown that restrictions on mobile sources would be cheaper alternatives than the retro-fitting additional pollution controls to existing power stations in those countries which have more sunlight, higher temperatures and less wind.

5.2   DAMAGE MECHANISMS

(a)   Deterioration of materials

Ozone is such a reactive gas towards organic molecules, be they part of plants, animals or useful materials, that it is worth examining the consequences in detail. Any double bonds in hydrocarbons are likely to be highly sensitive to chain-breaking and cross-linking reactions initiated by ozone. The likely mechanism of chain-cutting shown in Fig. 5.4 gives rise to peroxyl radicals which then can be photochemically excited to generate more free radicals in the same way that

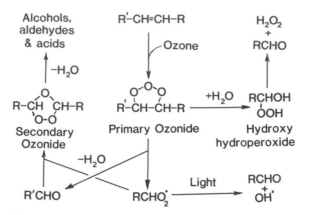

Fig. 5.4 The basic sequence of reactions associated with the process of ozonolysis. The main points to note are that the presence or absence of water dramatically affects the possible reactions and that, in a biological context where water is present, free radicals are not involved.

photochemical smog chain reactions are propagated (e.g. Reactions 5.16 and 5.17); the only difference is that they occur on the surface of materials that are being attacked causing loss of tensile strength and damage to the saturated components as well. Natural polymers like rubber, cotton, cellulose or leather as well as paints, elastomers (mainly used in tyre manufacture), plastics, nylon and fabric dyes are all degraded. Only when the double bonds are protected by adjacent groups can attack be resisted. The electronegative chlorine atom adjacent to the double bond in neoprene is a good example of this. Similar protective mechanisms have been built into more recent products but sacrificial auto-oxidants can be incorporated into others. The global economic costs of annual damage to materials due to ozone, and efforts to protect against it, are huge and constitute just over 30 % of that caused by all forms of atmospheric pollution to non-living materials.

(b) Ozonolysis or peroxidation?

Most materials in biological tissues are in close proximity to water. This changes quite considerably the characteristics of ozone attack upon unsaturated carbon compounds. Instead of free radicals and secondary ozonides being generated, hydrated forms of ozone may form hydrogen peroxide (see Fig. 5.4), as well as aldehydes, by a process called 'ozonolysis'. If the compound under attack by ozone has more than one double bond in the chain, for example in the fatty acid chains of lipids, then malonaldehyde ($OHC.CH_2.CHO$) may be

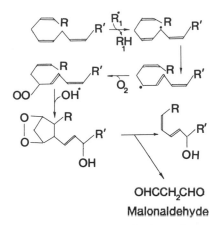

OHCCH$_2$CHO
Malonaldehyde

Fig. 5.5 Reaction sequence to illustrate the basic features of lipid peroxidation. Free radicals are involved both as intermediates and during the initiation process. Note that one of the products, malonaldehyde, is not unique to peroxidation but may also be produced during ozonolysis (Fig. 5.4).

produced (Reaction 5.19) which can easily be detected spectrophotometrically. The same compound may also be produced from similar substrates (e.g. fatty acids) by a different series of reactions known as (lipid) peroxidation. In the minds of some, ozonolysis and peroxidation are thought to be synonymous although they are quite distinct processes. The former creates hydrogen peroxide, involves no free (peroxyl) radicals in the initial attack and does not bring double bonds into conjugation with each other (i.e. produce alternate double and single carbon–carbon bonds). Lipid peroxidation, on the other hand, produces conjugated products with a strong spectrophotometric absorbance at a wavelength of 233 nm, involves initial attack by free radicals rather than by ozone itself and does not form hydrogen peroxide (Fig. 5.5). These distinctions between the two processes are therefore vital for an appreciation of which series of reactions and reactive species (ozone or free radicals) actually attack biological materials such as the proteins or lipids of membranes. Production of malonaldehyde by biological material is not evidence for one process instead of the other because it is produced by both. Even so, claims of lipid peroxidation are often made by those who do not understand these differences only on the evidence that malonaldehyde is formed during ozone attack. Other data obtained with some difficulty with

$$R.CH = CH.CH_2.CH = CH.R' \longrightarrow RCHO + OHC.CH_2.CHO + R'CHO \qquad (5.19)$$

model systems at very low temperatures show that free radicals from peroxidation could be involved in attacks on membranes but these temperatures are not likely in most biological materials. On the other hand, hydrogen peroxide formed during ozonolysis has been more readily detected.

It is possible that ozone may form hydroxyl radicals (Reaction 5.20) in alkaline solutions or by reacting with hydrogen peroxide (Reaction 5.21). Within living biological systems, however, hydrogen peroxide would be very quickly removed by catalase action (Reaction 5.22).

$$O_3 + H_2O \longrightarrow 2OH^\bullet + O_2 \qquad (5.20)$$

$$H_2O_2 + O_3 \longrightarrow HO_2^\bullet + OH^\bullet + O_2 \qquad (5.21)$$

$$2H_2O_2 \longrightarrow 2H_2O + O_2 \qquad (5.22)$$

There is less misunderstanding about the attack of low levels of oxides of nitrogen on similar unsaturated structures. These mechanisms involve the formation of free radicals as well as nitrous acid (Reaction 5.23) and show great similarity to those of (lipid) peroxidation. If this mechanism were the basic mechanism which gave rise to toxicity in both plants and animals then ozone and oxides of nitrogen would be equally toxic and produce similar types of damage. This is clearly not the case, ozone is much more harmful and must indicate a different type of attack by ozone to that of the oxides of nitrogen. Most probably, ozone attack takes place by ozonolysis as well as peroxidation because it may also react without producing free radicals although hydrogen peroxide is often identified as a side product.

$$NO_2 + R'.CH = CH.CH_2.R \longrightarrow HONO + R'.CH = CH.C^\bullet H.R \qquad (5.23)$$

(c) Protein sensitivity

Three amino-acids (cysteine, methionine and tryptophane) are sensitive to attack by ozone in aqueous solutions. The sulphydryl groups of the first two are oxidised to disulphide bridges or to sulphonate residues whilst the pyrrol ring of tryptophane is opened up to form *N*-formylkynurenine (Fig. 5.6). Similar reactions to free amino-acids within biological tissues would also be harmful but if they occurred to the same amino-acids forming part of proteins, which carry out critical functions within the cell, then the impact could be very much greater. These disturbances would be especially damaging if one or more of these amino-acids critically affected the secondary and tertiary shapes of a protein. Such changes in spatial orientation would be critical if, for example, they formed part of the reaction centre of

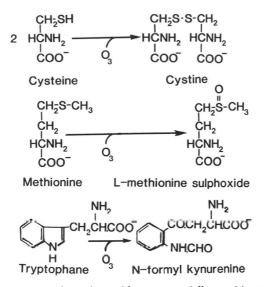

Fig. 5.6 Some groups in amino-acids are especially sensitive to oxidation by ozone. These include the sulphydryl groups of cysteine which form disulphide bridges, or the sulphur atoms of methionine which may be oxidised to sulphoxide, whilst tryptophane is very sensitive to opening of the pyrrol ring. All these reactions may take place with the free amino-acids or when they are already incorporated into proteins when greater damage may be caused.

an enzyme. Some enzymes certainly show clear evidence of both types of attack and there is little evidence that, once damaged by ozone, the original activity or shape of an enzyme may be regained. The proteins in plants, microbes and animals most likely to be exposed to ozone attack are those partially embedded within the cell membranes. The consequences of this type of membrane disruption are considered later.

(d) Peroxyacyl nitrates

Less research has been carried out to follow damage by peroxyacyl nitrates, not because the overall quantity of damage is less than that of ozone but because they are very difficult to generate in laboratory conditions. Peroxyacyl nitrates or PANs are capable of oxidising similar compounds to those known to be sensitive to ozone (e.g. methionine to methionine sulphoxide or cysteine to cystine; see Fig. 5.6) so the consequences to the thiol groups of proteins exposed to PANs are similar to those attacked by ozone: a change of structure and loss of reaction centre activity. Fortunately, the penetration of PANs into proteins is relatively poor and the half-life of most PANs is short (about 7 min at pH 7). Moreover, some key biological electron donors like NADH and NADPH (see Ch. 2) are only oxidised by PANs to their natural counterparts, NAD^+ or $NADP^+$ and are not destroyed as is the case with ozone. PANs, on the other hand, absorb a larger amount of the natural free-radical scavenging capability of biological tissues than ozone.

Normally, ozone causes the conversion of two molecules of reduced glutathione (a tripeptide called γ-glutamylcysteinylglycine often abbreviated to GSH) to a molecule of oxidised glutathione (abbreviated to GSSG) by virtue of the formation of disulphide bridges (Reaction 5.24). Peroxyacyl nitrates may also produce an additional thioester by reaction with three molecules of reduced glutathione (Reaction 5.25) which cannot be easily replaced by natural glutathione regenerating mechanisms (glutathione reductase, etc., see below). This effectively reduces the working pool size of natural free-radical scavengers and, as a consequence, more damage may be done.

$$2GSH + O_3 \longrightarrow GSSG + H_2O_2 \tag{5.24}$$

$$3GSH + CH_3CO.O_2NO_2 \longrightarrow GSSG + GS.CO.CH_3 \tag{5.25}$$
$$+ H_2O + HONO_2$$

(e) Natural free-radical scavenging

Apart from molecular oxygen or ozone or atomic oxygen, successive univalent reductions of molecular oxygen may give rise to superoxide radicals ($O_2^{\cdot-}$), hydrogen peroxide, hydroxyl radicals and water (Fig. 5.7). Superoxide and hydrogen peroxide are not themselves sufficiently reactive in aqueous solution to initiate lipid peroxidation but the protonated form of superoxide, the perhydroxyl radical ($HO_2^{\cdot-}$) can attack membrane lipids. The pK_a of the equilibrium (Reaction 5.26) at 4.8 is relatively low so in well buffered cell contents (around pH 7) only a few perhydroxyl radicals may exist. In cell vacuoles, however, where acidity levels are higher, the situation may be more conducive to their formation. It may also be the case that unbuffered extracellular fluid layers experience relatively high acidity levels due to simultaneous exposures to atmospheric quantities of sulphur dioxide and nitrogen dioxide which also favour the formation of perhydroxyl radicals at the expense of superoxide (see Ch. 7).

$$H^+ + O_2^{\cdot-} \rightleftharpoons HO_2^{\cdot-} \tag{5.26}$$

The relative toxicity of superoxide as compared with hydroxyl radicals has been the subject of much discussion and investigation

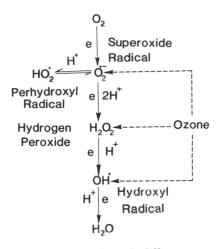

Fig. 5.7 The reduction of oxygen through different stages to illustrate the relationship between the different forms. Note especially that ozone may form a number of different reactive states and that the superoxide radical is in equilibrium with the perhydroxyl radical in a one to one relationship at pH 4.7.

(see Appendix, Sect. 8.2, for a full description). Any cellular environment in which such a free radical exists greatly affects the reactivity of this free radical. In an aqueous environment, it is relatively unreactive but in the hydrophobic regions of the fatty acid side chains inside membranes it oxidises the natural antioxidant α-tocopherol (vitamin E) and removes one of the most important non-aqueous antioxidant mechanisms in a cell (see below).

Free hydroxyl radicals on the other hand react with virtually all biological molecules almost immediately so they are not able to diffuse any distance without reacting. The two are linked by the different reactions that convert superoxide into hydroxyl radicals (e.g. Reactions 5.27 to 5.29), where Reaction 5.28 (catalysed by the enzyme xanthine oxidase) is shown here only as an example. There are many other similar mechanisms by which release of hydrogen peroxide into biological tissues may take place.

$$O_2^{\cdot -} + Fe^{3+} \longrightarrow Fe^{2+} + O_2 \qquad (5.27)$$

$$\text{hypoxanthine} + O_2 \longrightarrow \text{xanthine} + H_2O_2 \qquad (5.28)$$
$$\text{(or xanthine)} \qquad\qquad\quad \text{(or urate)}$$

$$Fe^{2+} + H_2O_2 \longrightarrow Fe^{3+} + OH^- + OH^{\cdot} \qquad (5.29)$$

To avoid the hydroxyl radical from being formed, all biological tissues have a series of anti-oxidant defences. Some are specific to certain species but a great number occur ubiquitously in all aerobic (oxygen-requiring) organisms. Superoxide dismutases, for example, are metalloproteins which catalyse Reaction 5.30 and are of three basic types: a copper–zinc protein common to vertebrates including humans, higher plants and fungi, a manganese enzyme in all animal and plant mitochondria as well as bacteria, and an iron-containing superoxide dismutase only found in certain prokaryotic organisms (i.e. those without distinct membrane-bounded compartments). To operate properly they require the cooperation of another widely distributed enzyme (catalase) to break down the hydrogen peroxide (Reaction 5.22) which would otherwise form hydroxyl radicals (Reaction 5.29).

$$2O_2^{\cdot-} + 2H^+ \longrightarrow H_2O_2 + O_2 \tag{5.30}$$

At least three other lines of anti-oxidant defence are common to most organisms, all of which may be regarded as semi-sacrificial. Glutathione peroxidase, for example, may remove hydrogen peroxide and other organic peroxides from aqueous parts of the cells by catalysing the conversion of two molecules of glutathione (GSH, see Sect. 5:2d) to one oxidised molecule of glutathione (GSSG, Reaction 5.31). Reduced glutathione is then regenerated by another enzyme, glutathione reductase by a different mechanism.

$$2GSH + ROOH \longrightarrow GSSG + H_2O + ROH \tag{5.31}$$

Vitamin E (α-tocopherol) is formed by plants but is required in the diet of humans mainly as an essential antioxidant. In both animals and plants, it is oxidised first to a semiquinol and then to tocopheryl quinone by superoxide radicals, hydrogen peroxide or hydroxyl radicals (Fig. 5.8). Being naturally lipophilic (or hydrophobic), it appears in cellular locations especially membranes near to sensitive components (such as the electron transport intermediates of plastids or mitochondria: see Appendix, Sects. 8.3 and 8.4) which are especially susceptible to free-radical formation and difficult for hydrophilic antioxidants to approach. Some tocopheryl quinone may be reduced back to α-tocopherol but sacrificial losses are often considerable and must be replaced by fresh synthesis in plants or through the diet of humans if harmful oxidations within membranes are to be avoided.

Vitamin C (ascorbate) is another anti-oxidant made by plants, taken up in the diet of humans and used in a similar scavenging role in both animals and plants. This carbohydrate may react with superoxide radicals, hydrogen peroxide or hydroxyl radicals to form dehydro-ascorbate. This may be catalysed by at least two different enzymes, ascorbate oxidase or ascorbate peroxidase (Fig. 5.8). In both, a mono-dehydro-ascorbate free-radical intermediate may be formed. Enzymes to reduce both the mono- and di-dehydro-ascorbate back to

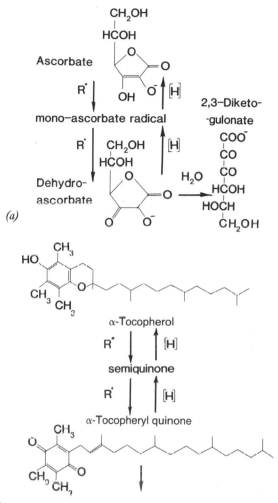

(a)

(b)   Oxidation to other metabolites

Fig. 5.8(a) and 5.8(b)   Two important free-radical scavenging systems found in most biological systems include the participation of the anti-oxidants ascorbate (vitamin C) and α-tocopherol (vitamin E). In both systems free radicals cause the oxidation of the anti-oxidants in a two-stage process and the oxidised products are partially recovered by separate reduction processes. Ascorbate is especially effective in scavenging free radicals in aqueous environments whilst α-tocopherol protects sensitive elements in non-aqueous (hydrophobic) regions especially in bioenergetic membranes where free radicals are frequently produced and also where they can do most damage.

ascorbate are universally present but inevitably some ascorbate is lost as the sugar ring opens to form 2,3-diketogulonate which is then metabolised away. Constant replenishment of ascorbate by recycling, by fresh biosynthesis in plants or from restricted reserves in animals is continuously required to maintain this essential scavenging activity in the aqueous region of cells.

Consequently, because ozone may readily form a variety of different free radicals (Fig. 5.7), any consideration of attack upon different biological systems must therefore always take these natural protective mechanisms into account. Moreover, the nature of the environment within the cell determines which antioxidant is consumed. Aqueous phases, favour the use of glutathione and ascorbate, non-aqueous (or hydrophobic or lipophilic) regions are more dependent on the protection offered by α-tocopherol.

5.3 PLANTS

(a) Access

Deposition velocities of ozone and peroxyacyl nitrates from the atmosphere to various surfaces on the Earth (soil, water and vegetation) are higher than those of the oxides of nitrogen and about the same as those recorded for sulphur dioxide (Table 3.2). Still air conditions during photochemical smog episodes provide partial protection to vegetation because of the undisturbed boundary layers above and below the leaves which offer some resistance to the diffusion of ozone. Later, as winds move the tropospheric ozone around, damage to vegetation may be experienced quite a distance away from the locality in which they were generated because the boundary layers of these leaves are stripped away by moving air.

Access of ozone (and peroxyacyl nitrates) into the leaf is now known to be mainly through the stomata, although waxy leaf surfaces (cuticles) may be partially degraded by ozone especially any of the unsaturated elements within the wax. Fortunately, chemical changes within the wax are rare in most species and concentrations of pollutants have to be especially high or prolonged for this to occur although physical changes in the structural organisation of waxes are more frequently detected. Most plants have open stomata during the day, so damage is largely confined to this period but in dry regions a proportion of the plants (e.g. cacti) only open their stomata at night to conserve water loss. Their period of maximum sensitivity to ozone is therefore during periods of darkness.

It is thought by many that the degree of stomatal opening is the principal, or even the sole, response in determining differences in

Table 5.1   Factors that influence plant response to ozone and PANs

Genetic	Environmental
Developmental (and ageing)	Soil conditions
Cell	Acidity
Tissue	Temperature
Leaf	Water stress
Root	Nutrients[†]
Plant	
	Climate conditions
	Light
Between species	Humidity
	Temperature
	CO_2 levels
Within species	Other pollutants
Individuals	
Clones	Biotic conditions
Varieties	Pathogens
Cultivars	Individual competition
Populations	Species competition

[†] Especially potassium and nitrate.

plant response to ozone and peroxyacyl nitrates. These responses may be of two types, environmental or genetic (Table 5.1). But the assumption that flux of ozone into leaf tissue is inversely proportional to stomatal resistance may not be entirely correct and other factors may be involved as well. Moist surfaces within the leaves (e.g. the extracellular fluid of mesophyll tissues) allow the ozone to dissolve and diffuse down a concentration gradient similar to that of carbon dioxide. Solubility, rate of decomposition and pH of the medium all influence the amount of ozone taken up and passed onwards into the plant tissues. Ozone is approximately one third as soluble as carbon dioxide so the correlation between them in terms of pathway flux is probably valid but it must be remembered that it is over a hundredfold less soluble than sulphur dioxide. This means that sulphur dioxide uptake takes place mainly on the lower inner surfaces of the stomatal guard cells, where most (77–90 %) of the transpired water is lost, rather than from the mesophyll extracellular fluid which contributes only 10–23 % of water loss but 85 % of the uptake of carbon dioxide (see also Fig. 2.10). This is largely because removal mechanisms of carbon dioxide are associated with photosynthetic carbon fixation. Ozone uptake also has a similar pathway to carbon dioxide because of the sink effect of the mesophyll membranes. The products of sulphur dioxide, by contrast, are accumulated to a much greater extent before removal and therefore follow the pathway of bulk water flow in reverse.

(b) Cellular changes and damage

The cell wall (mainly cellulose) offers little barrier to diffusion of ozone nor is it attacked to any significant extent by ozone or the peroxyacyl nitrates. Once in solution, however, ozone immediately forms other derivatives which show varying degrees of reactivity. Amongst these must be included hydroxyl radicals, monatomic oxygen, hydrogen peroxide or superoxide radicals ($O_2^{\bullet-}$) and it is these which may be actually responsible for most of the harmful effects of ozone (Sect. 5.2d). The membrane proteins and unsaturated regions of the fatty acids within membranes are readily attacked by ozone and peroxyacyl nitrates by mechanisms already outlined (Sec. 5.2) as well as by these harmful free radicals. This may occur before the natural free-radical scavenging systems (Sect. 5.2d) within the cells can counteract the effects of some of them. Most biochemical, electron microscopical and electrophysiological studies of ozone damage have indicated that the plasma or cell membrane of plant cells (sometimes called the 'plasmalemma') suffers the most injury. This is characterised by changes in permeability and 'leakiness' of cell membranes to certain important cations like potassium. Other cellular membranes, such as the envelopes of the mitochondria or chloroplasts, are similarly damaged but to a lesser extent as the effect of ozone and its toxic radicals are progressively diluted and absorbed from the outside inwards. This partially explains the disturbance of ultrastructure within both these organelles and inhibition of both photosynthesis and respiration caused by high doses of ozone. There is some evidence of cellular repair but this consumes energy which is in short supply because of damage to bioenergetic organelles. Exact sites of injury are also hard to locate, unlike those caused by sulphur dioxide, because of the more random nature of ozone attack upon unsaturated groups and sulphydryl residues. Disturbed functions within plants due to ozone are supported by losses of chlorophyll, increases in leaf fluorescence (indicative of wasted light energy) and changes in adenylate (ATP, etc.) levels.

The best indications of perturbation within plant cells due to realistic exposures to ozone are shown by changes in the fluxes of various components across membranes, especially those of sugars, amino-acids, water and potassium ions. When the levels of ozone are low, or only present for short periods, recovery of normal transport functions is possible. Moreover, the nature of this ozone damage appears to be associated with the specific sites of transport rather than being spread out across the whole of the membrane area. As these particular regions consist of proteins rather than lipids, this is taken to mean that ozone attacks the sulphydryl groups of proteins rather than the double bonds of lipids by the process of ozonolysis (see Sect. 5.2b). The consequences of disturbed transport across membranes are indicated by reductions in the ability of the cell to control osmotic pressure and to maintain effective electropotentials across the plasma membranes and, to a lesser extent, across energy-forming mitochondrial and chloroplastidic membranes. Most recent observations would suggest that it is specific damage by ozone to those plasma membranes engaged in the heaviest transport which are the most critical events. This would involve the plasma membranes of the stomatal guard cells or those cells engaged in loading the transporting or vascular elements of leaves which are called companion cells in flowering plants (angiosperms) or Strasburger cells in conifers (gymnosperms). It also means that increased levels of acidity and other ions (e.g. sulphate) might be expected to occur at these damaged locations and together they would have a cumulative but highly detrimental effect upon the ability of a plant to move and use water and photosynthate.

(c) Visible injury

Injury by ozone is generally the consequence of an inability to repair or compensate for altered membrane permeability. Initially confined to sensitive cells, the net effect of prolonged ozone attack leads to irreversible pathological damage characterised by bleached areas. The first symptoms of such injury are shown by water-soaked areas of leaf due to leakage of cells just below the epidermis. If recovery is still possible these shiny water soaked regions may disappear as the tissues repair themselves and recover their permeability control.

An added complication to such injury is that some ozone-sensitive species and cultivars tend to produce more ethylene in the presence of ozone as this is one of the primary responses of a plant to stress caused by a variety of unfavourable environmental conditions. This gas itself is a potent plant growth regulator which promotes leaf fall and early senescence. Ethylene is produced almost as soon as permeability changes occur and will decline if recovery is made. By then, however, some of the ethylene-induced consequences may have been promoted so, even if no permanent visible injury appears, the recovered leaves will have a shorter life thereafter.

Although visible injury is normally confined to the green foliage regions of plants, a variety of other environmental factors may influence the type and nature of visible injury or mimic the effect of ozone (Table 5.2). The usual type of damage is seen as chlorotic flecking

Table 5.2 Other factors which influence the extent of ozone-associated damage to plants in addition to those listed in Table 5.1

---

1. Physical characteristics of leaves
   (a) Cuticular wax
   (b) Hairs
   (c) Stomatal structure
2. Vigour
   (a) Leaves
   (b) Roots including mycorrhizal associations
3. Degree of succulence
4. Age
   (a) Whole plants
   (b) Leaf
5. Exposure to ozone
   (a) Length
   (b) Concentration
6. Exposure to PANs and other air pollutants
7. Environmental (see Table 5.1).
8. Genetic (see Table 5.1)
9. Associated injury from other biotic factors
   (a) Insects
   (b) Fungi
   (c) Bacteria
   (d) Viruses
   (e) Pesticides

---

or stippling of the leaves between the veins but this type of this damage varies tremendously between species. Such injuries, mainly confined to the palisade cells of the upper mesophyll, may become reddish or bronzed due to enhanced anthocyanin production or give a darker appearance to the leaf due to the stimulation of tannin production.

(d) Peroxyacyl nitrate damage

PANs and other similar substances enter a leaf through the stomata and may be as toxic to plants as ozone. Visible injury symptoms differ from those of ozone, with bronzing of the lower surfaces of leaves being a typical feature. Developing regions of leaves are the most sensitive to PANs and this feature may give 'stripe-like' damage to the leaf when a smog episode coincided with growth of the leaf at that point. By contrast, on either side of this region, the adjacent leaf tissue may appear quite healthy. The targets of damage are again the sulphydryl groups of protein and, to a lesser extent, the unsaturated double bonds of lipids (see Sect. 5.2c). For such damage to plants to have occurred there must have been illumination before, during and after exposure to PANs. It could be that proteins, which are responsible for modulating metabolic activity due to the presence or absence of light are more sensitive or, alternatively, critical events in the light reactions of photosynthesis are even more susceptible to damage by PANs than by ozone. A more likely explanation is that the natural mechanisms of free-radical scavenging (i.e. those described in Sect. 5.2d) are interrupted or interfered with as they cope with normal photo-oxidations in the light. In darkness they have more than sufficient capacity to cope with PANs and immediately allow repair. Exposure of plant tissues to PANs in the light comes at a most critical time when free-radical scavenging mechanisms are stretched to their limit and hence unrepaired damage will deteriorate as the products of photo-oxidation build up. This, in turn, starts fresh and uncontrolled free-radical-initiated disturbances. The 'stripe-like' damage to developing tissues caused by PANs is consistent with this type of damage. These regions have the greatest sensitivity to photo-oxidation because they are still trying to grow. This means that a critical balance is disturbed between diversion of energy and growth materials for construction purposes and similar expenditures required for normal maintenance (which includes free-radical scavenging and repair). If PAN exposure episodes coincide with phases of rapid or sensitive growth then the developmental process for the tissues of that region at that time is disrupted. Thereafter, natural development for the leaf is continued when conditions improve leaving behind the injured area as a damaged band between healthy regions.

5.4 ANIMALS

(a) Hazards at work

Problems due to ozone pollution indoors are often unappreciated by most people and the situation could get worse. In the past there was some appreciation by those working with specialised equipment such as X-ray machines, ultraviolet lamps, xenon arc lighting and other high-voltage generating equipment, that one of the many associated hazards was the release of ozone due to ionising influence of the electrical discharges involved. Perhaps the most hazardous task involving exposure to ozone over a long time has been electric-torch welding which is often carried out in restricted spaces. In such cases it is normal practice to take care that adequate ventilation is provided so as to reduce the problem. Similarly, in aircraft flying at very high altitude, account must be taken of the much higher levels of ozone in the stratosphere and adequate protection should be provided for pilots. However, the advent of photocopiers, their huge popularity and their disposition throughout many workplaces not originally designed for such equipment has brought more instances of unacceptable ozone levels indoors. Repetitive noise in the office usually causes such machines to be semi-isolated to poorly ventilated corners or small side rooms, and the problem is made worse.

Recommended working levels (TLVs or OELs) of ozone indoors during a 40-hour working week of below 0.1 ppmv ozone are probably still too high and levels indoors well below 0.05 ppmv are more desirable. Ambient levels outdoors are usually around 0.005 ppmv, but lightning within thunderstorms or photochemical smog episodes may raise levels well above 0.1 ppmv. The United States standard of a hourly mean of 0.12 ppmv ozone outdoors not to be exceeded more than 1 day per year, is actually exceeded many times during summer months in the USA, Europe and Japan. High-risk groups include children, the aged, individuals with pre-existing respiratory or immune-related diseases, people with coronary heart disease and those with glucose-6-phosphate dehydrogenase deficiency syndrome. With ozone it is the concentration rather than the length of exposure that has more effect (i.e. a short exposure to high concentration causes proportionally more harm than a long exposure to low levels of ozone) and intermittent exposures are worse than continuous exposures (i.e. 6 hours of ozone and clean air for the rest of the day is as bad as ozone-polluted air at a similar concentration for the whole day). If other pollutants are also present (usually oxides of nitrogen) then the possibility that the effects may be synergistic (i.e. more than additive) is greater.

(b)   Short- and long-term exposures

Photochemical smog causes irritation to the eyes, nose, throat and chest. The classic eye irritation caused by such conditions is caused not so much by ozone as by peroxyacyl nitrates and trace free-radical hydrocarbon fragments. Unfortunately, little has been done to separate distinctive medical responses due to these agents from those of ozone, although an immense amount of study has been devoted to determining the conditions under which ozone exposure causes tumours. Once again, this has mainly been done by using animals as model systems with all the attendant problems of translating observed animal responses into likely human consequences (see Chs. 2 and 3) and differences of interpretation remain.

There is general agreement, however, that ozone is a powerful oxidant which may injure the bronchiolar and alveolar walls of the lungs. Surface epithelial cells are damaged or destroyed which later may be replaced by thick cuboidal cells with few or little cilia (cell hairs). This initial damage may be accompanied by an accumulation of water within the affected tissues (oedema) if the levels of ozone are especially high, and together these symptoms may give rise to an acute inflammatory response. This only occurs at high levels of exposure, diminishes with reduced concentrations of gas but is made worse by exercise. Ultrastructural changes to the epithelial (surface lining) cells during such attacks include the loss of cilia, cytoplasmic vacuolation (formation of internal cellular spaces) and condensation of mitochondria into larger structures with abnormal cristae. Similar initial damage at the cellular level may be caused by nitrogen dioxide but at twenty times the concentration of ozone – thus illustrating the reactivity of this particular photochemical oxidant.

Over long periods of exposure, replacement of epithelial cells within the lungs may be considerable and there is an increase in the number of macrophage cells (certain white blood corpuscles), fibrous elements, and mucous-secreting cells as well as a thickening of the walls of the major airways. These may exacerbate a number of clinical problems such as bronchitis and emphysema. On the other hand, enhanced tolerance to subsequent exposures to ozone has been reported in animals and humans after an initial exposure. Higher initial concentrations of ozone have been found to give rats more protection against re-injury than lower initial levels of ozone when both groups were subsequently re-exposed to double the initial concentrations. Similarly, when biochemical changes in the blood of California residents were compared to those of Canadians exposed to similar concentrations of ozone, a number of indicator tests showed that some tolerance had already been acquired by the Californians. Epidemiological studies of enhanced respiratory infections and reduced lung function in populations exposed to photochemical smogs have been hampered by the fact that such surveys are dealing with complex and changing mixtures of ozone as well as oxides of nitrogen, PANs and other components of smog. Shorter-term dose-response studies have indicated that although oxides of nitrogen and ozone cause similar adverse effects on lungs, more than five times the level of nitrogen dioxide are required to elicit similar detrimental effects to those caused by ozone.

Tolerance mechanisms have been studied extensively by toxicologists, and ozone is a particularly good inducing agent to elicit this phenomenon. Moreover, it initiates a variety of cross-tolerance reactions which occur when one toxic substrate causes the acquisition of

Table 5.3 Positive cross-tolerance relationships

Primary agent	Secondary agents which induce tolerance to the primary agent
H_2S	O_3, $COCl_2$
NO_2	CCl_3NO_2, O_3, $COCl_2$, thiourea
O_3	CCl_3NO_2, H_2S, ketene, NO_2, NOCl, $COCl_2$, thiourea

tolerance to another irritant. A number of these possibilities are known and their relationships to some of the other known harmful air contaminants are shown in Table 5.3.

(c) Biochemical and physiological changes

Animals have a wide range of protective free-radical scavenging systems (see Sect. 5.2d) and exposure to ozone may cause increases in levels of supcroxide dismutase, glutathione peroxidase, glutathione reductase, disulphide reductase and non-protein sulphydryl groups (mainly glutathione) within lung tissues, which may in turn account for some of the acquired tolerance over a period of time. Levels and rates of consumption of ascorbate have not been intensively studied but animal tissues which are adequately supplied with vitamin E (α-tocopherol) are much less sensitive than those known to be deficient in this protective compound. Increased production of malonaldehyde is indicative of more ozonolysis rather than peroxidation (see Sects. 5.2b and 5.2c) under these circumstances.

Disturbances of mitochondrial structure, reduction in lung tidal volume and increases in rates of respiration have caused a number of biochemical parameters to be followed by clinical scientists. These studies have shown that, whilst mitochondrial preparations can be induced to show a variety of changes (i.e. initial swelling, increased levels of enzymes such as NADH–cytochrome c reductase or succinate oxidase activities, enhanced oxygen consumption and uncoupling – see Appendix, Sect. 8.4) it is not known if the mitochondria in the replacement cuboidal cells or invading macrophages are more active or whether the mitochondria of the original tissues are altered after exposure. Explanations based on precise disturbances within mitochondria do not appear to be satisfactory answers given that ozone is known to produce a range of free radicals (see Sect. 5.2) which can influence membrane-linked mitochondrial functions (i.e. transport, permeability or 'leakiness') in a variety of different ways – very similar to those that occur in sensitive plant tissues (Sect. 5.3b).

Maintaining sterility of lung tissues is one of the major functions of the alveolar macrophages. These white blood cells, derived from the bone marrow, mainly operate by engulfing invading particles (including bacteria) and taking them into a protective vacuole inside themselves which is often called a 'phagosome'. Enzymes such as acid phosphatase and lysozyme are delivered into this vacuole by a process known as 'degranulation' associated with other organelles called 'lysozomes'. The hydrolytic processes that follow break down the contents but the vacuolar membrane protects the rest of the macrophage from these powerful breakdown enzymes as they work upon and destroy the incapacitated invaders. Normally, the hydrolytic enzymes of the macrophages are carefully packed away in specially resistant organelles called 'lysozomes' to avoid harm being done to the rest of the macrophage but ozone is capable of causing increased fragility and partial lysis of these lysozomes. Lysozomal enzymes are normally used to destroy other cells and other ingested particles within phagosomes but upon death of the cell itself they are also released upon their own internal structure to bring about self-digestion which is a normal event associated with death. Ozone thus has the effect of initiating this possibility before the normal lifespan of a cell is completed. Furthermore, exposure of lung tissues to ozone appears to induce such macrophages to congregate and increase their activities just as though bacterial invasion was being experienced even though no infection has started. More macrophages are drawn in as though a major bacterial attack was in progress and they become partially disabled depending on the degree of exposure to ozone. At the same time, their original job of fighting bacterial infection is partially suspended so that pneumonia and other respiratory illnesses take hold more easily. Combinations of effects by ozone are therefore brought about and made much worse by slight but highly significant disturbances of cellular activity which normally exists for other purposes. It also provides an insight into how relatively low levels of ozone can bring about significant changes in lung function.

(d) Cellular models

Red blood cells (erythrocytes) carry vast amounts of oxygen around the body and to do this the cell membranes are freely permeable to non-reactive dissolved gases. Oxygen is much more soluble in the hydrophobic (water-repelling) regions of membranes but ozone is ten times more soluble than oxygen in aqueous media such as the blood plasma and in the cytoplasm of erythrocytes. Consequently, human red blood cells (erythrocytes) have been much used as model cells to study effects of ozone on cell membranes. These particular cells also have the advantage that they do not carry out protein synthesis after they have lost their nuclei and only metabolise sugars using glycolytic and associated pathways. As a consequence, inherent adaptive biochemical processes which might occur during experimental exposure are not a complication. Such studies have revealed numerous sites of sensitivity to ozone and its products, hydrogen peroxide, both on the outside and inside of the cell by virtue of changes in membrane fragility, depressed levels of anti-oxidants such as glutathione and changes in enzymic activity of certain enzymes.

Mechanisms of ozone attack on individual proteins have been examined in great detail, especially those on glycophorin (a trans-membrane protein) which is inserted across erythrocyte cell membranes. Amino-acids numbered 72 to 92 in sequence of this protein (from the N-terminal amino-acid) are embedded within the membrane and do not contain cysteine or tryptophane amongst their number. Methionine at number 8 is to be found on the outside of the red blood cell whilst that at position 82 is inside the membrane. If ozone is allowed to attack this protein when it is embedded in artificial lipid vesicles, methionine 8 is oxidised to methionine sulphoxide (Fig. 5.6) but methionine 82 is protected. Changes like this explain why immune responses, which depend on exposed surface proteins, are also altered by ozone because it (or its derivatives) react more rapidly with exposed methionine, cysteine and tryptophane residues on the outside of cells. If the levels of ozone exposure are increased or prolonged then other changes to sensitive sulphydryl-dependent proteins beyond the cell membrane (e.g. the cysteines of glyceraldehyde-3-phosphate dehydrogenase) or even beyond two membranes (e.g. the tryptophanes of lysozyme within the lysozomes) may also show similar distinctive oxidations at specific sites.

Other experiments with red blood cells have demonstrated another important point. Levels of ozone have to be much higher for effects on lipids to occur than those required to cause changes in trans-membrane proteins such as glycophorin. Addition of a lysophospho-lipid to red blood cells causes cell rupture or 'lysis' (as is commonly caused by the venom of snake bites or insect stings) whilst exposure of red blood cells independently to ozone does not cause lysis. Treatment of normal phospholipids with ozone in the absence of red blood cells also causes the fatty acid chains to break at their double bonds by the process of ozonolysis (i.e. no free radicals are formed). These ozone-modified phospholipids in the absence of ozone then behave as if they were lysophospholipids and are subsequently able to lyse or break open red blood cells. This means that the ozone at these concentrations cannot generate ozonised phospholipid whilst it is buried within the cell membrane yet it can change exposed sulphydryl groups both on the outside and within cell membranes. Therefore any major disruption by ozone upon membranes, and the cellular functions that depend upon membranes, is almost certainly initiated by ozone attack on the proteins of the membrane rather than the lipids although this primary damage within the proteins may lead subsequently to secondary disturbance of lipid structures.

(e) Mutagenesis

Ultraviolet light and ozone are commonly used to kill microorganisms in swimming pools when the use of hypochlorite needs to be avoided or reduced. The ozone appears to oxidise not only the sulphydryl groups and ozonate the fatty acid residues of the ubiquitous *Escherichia coli* and other undesirable bacteria, but also to participate with the ultraviolet light to cause genetic damage. This occurs by a variety of mechanisms including the physical rupture of the strands of DNA or by damaging the DNA repair mechanisms. It is thought that most of the chromosomal disruptions are actually caused by the ultraviolet light but the additional presence of ozone inhibits the DNA recovery mechanisms which are normally induced after ultraviolet light-initiated damage is detected.

Extensive tests using animals have been carried out to evaluate the mutagenic risks to humans directly due to atmospheric ozone. Very little evidence has been produced to support such a possibility but similar exposure studies using irradiated smog from car exhausts have produced, for example, a marked decrease in number of mice per litter, the frequency of litters per mother and survival rate of infant mice. The elevated mortality rate was traced to alterations in the genetic composition of the sperm which must have been caused by components other than ozone in the smog. Over the past decade it was particularly noticeable that despite such disturbing observations, studies of the effects of active agents of photochemical smog on animals and plants have gone out of fashion and yet the need to evaluate them properly has never been so great. At the moment the World Health Organization does not believe a guideline level for peroxyacyl nitrates lower than current TLVs (or OELs) is warranted because it does not appear to be a significant health problem. The long-term health hazards of many other active ingredients of photochemical smog and trace hydrocarbons other than those that also occur in cigarette smoke have still to be evaluated.

SELECTED BIBLIOGRAPHY

Adams, R. M., Hamilton, S. A. and McCarl, B. A., *The Economic Effects of Ozone on Agriculture.* US Environmental Protection Agency, 1984, Corvallis, Oreg.

Dotto, L. and Schiff, M., *The Ozone War.* Doubleday 1978, New York.

Grennfelt, P. (ed.), *Ozone – The Evaluation and Assessment of the Effects of Photochemical Oxidants on Human Health, Agricultural Crops, Forestry, Materials and Visibility.* Swedish Environment Research Institute, 1984, Gothemburg.

Guderian, R. and Rabe, R., *Photochemical Oxidants – Formation, Control, Effects on Man, Animals and Plants.* Springer-Verlag, 1983, Berlin, Heidelberg and New York.

Tucker, A., Air of uncertainty. *The Guardian*, 25 March, 1975.

VDI-Fachdokumentation Reinhaltung der Left, *Ozon und Begleitsubstanzen in Photochemischen Smog.*, 1970–1976, Dusseldorf.

6

Other pollutants – global and local

'Above the smoke and stir of this dim spot,
which men call Earth.'

From 'Comus' by John Milton (1608–1674)

6.1   OXYGEN?

(a)   Evolution of the atmosphere

If, according to Mellanby, a pollutant is 'a chemical in the wrong
place and at the wrong concentration' then, in planetary and stellar
terms, present levels of oxygen in our atmosphere are very unusual
and could qualify under this definition. Perhaps 'at the wrong time'
should be added to disqualify oxygen. The current concentration of
21 % oxygen makes the atmosphere of the Earth potentially explosive,
almost midway between 16 % oxygen where combustion is not possible and 25 % oxygen where spontaneous combustion would take
place – yet at a vital level for life as we know it. By contrast, oxygen
encourages a number of free-radical reactions within living cells and,
as a consequence, is still one of the major external problems encountered by biological systems.

In the early formative years, the atmosphere of the Earth consisted
mainly of carbon dioxide. Free oxygen only appeared in this atmosphere about 2 billion years ago as photosynthetic organisms acquired

the ability to split water into oxygen, electrons and protons. About 1 billion years ago, oxygen still formed only 1 % of the atmosphere but levels of carbon dioxide declined rapidly as it was fixed by photosynthetic organisms. These carbonaceous products passed into food chains and eventually ended up as carbonates in mollusc shells or corals. The consequence today is that most of the carbon dioxide from the original atmosphere of the Earth is now locked away in the limestone reserves of the world. Oxygen (the most abundant element of the Earth at 53.8 %) has thus been steadily released into the atmosphere to become toxic or, in most cases, lethal to many of the more primitive organisms that existed then because of harmful oxidations within their cells.

(b) Anaerobes versus aerobes

Organisms that developed protective mechanisms to cope with the appearance of oxygen survived and those that did not either died or withdrew to specialised and restricted environments where the oxygen concentration was especially low. Antecedents of the latter are represented today by a wide array of anaerobic organisms ranging from very 'strict' anaerobes that cannot tolerate oxygen at all, others which will survive exposure for some time but do not grow in the presence of oxygen, and 'moderate' anaerobes that grow in concentrations of up to 10 % oxygen. Consequently, there are whole ranges of sensitivity towards oxygen shown by different microbes as well as a large number of specialised environments in which oxygen levels are low and permit anaerobes to thrive. Undisturbed or stagnant soils, polluted waters, decaying vegetation, wounds and internal passages of animals (especially the mouth and the colon) are all locations where anaerobes flourish.

These localised environments must also be able to provide a ready supply of reduced compounds for provision of energy by means of biological oxidation and a flow of electrons through redox couples more electronegative than oxygen and water. The nitrate respiration carried out by anaerobic denitrifying bacteria (Sect. 3.1 and Fig. 3.4) is a particularly good example of this feature. In some instances, certain enzyme complexes are strongly inhibited by oxidation at their active sites. The nitrogenase complex (see Sect. 3.1 and Fig 3.3) is especially sensitive to oxygen and most aerobic organisms have special structures which isolate their nitrogen-fixing complexes from oxygen. For example, blue-green algae (or, more correctly, cyanobacteria) carry out nitrogen fixation in separate thick-walled cells (or heterocysts) whilst the root nodules of legumes have leghaemoglobin to bind up any oxygen that may appear. Other oxygen-limitation mechanisms include the possession of very low respiration rates or the capacity to turn off photosynthesis when nitrogen fixation takes place.

Aerobic organisms that have evolved in the presence of rising atmospheric oxygen concentrations over the past 2 billion years have extensive free-radical scavenging systems (e.g. vitamin E, glutathione reductase, etc., see Ch.5 and Appendix, Sect. 8.2) and some have acquired the use of oxygen as a terminal electron acceptor for respiration or oxidative phosphorylation. This has enabled more efficient oxidation of reduced substances to provide energy. All modern higher plants and animals (but only some microbes) have acquired this aerobic capability at some time during evolution. Usually, the term 'aerobic' is only applied to the aerobic bacteria with the assumption that it follows without statement that plants and animals are aerobes. However, there are a number of examples of the latter where respiration has been lost. Many nematodes do not require oxygen and there is even a fish, the crucian carp that completely cuts off normal respiratory activity for long periods of time when the lakes in which it lives dry out. Nevertheless, all plants, microbes and animals are damaged or become less efficient at concentrations greater than 21 % oxygen. There are many reasons to account for this inefficiency.

(c)   Photorespiration in plants

Most plants growing at levels of 21 % oxygen are inefficient by comparison with those grown at 1 % oxygen or in various concentrations of carbon dioxide (Fig. 6.1). Net fixation of carbon dioxide by plants may be improved by a factor of 50 % or more by reducing oxygen levels below 21 % but, if higher concentrations of oxygen than ambient are experienced, severe reductions of photosynthesis (Fig. 6.1), inhibited root growth, poor seed viability and stunted shoot development may also take place because additional free radicals give rise to uncontrolled damage in cellular membranes. Figure 6.1 also illustrates another significant feature which has only been fully appreciated in the last few years. The intercepts on the horizontal axis of the relationship shown represent a balance when carbon dioxide uptake matches the evolution of carbon dioxide in the light and are known as the compensation points. These vary with the partial pressure of oxygen: the more oxygen the higher the compensation point concentrations. It has been known for some time that only a portion of the evolution of carbon dioxide is independent of light and is mainly due to respiratory activity in mitochondria but the light-dependent remainder is due to a process known as 'photorespiration'. This would initially appear to be a futile waste of photosynthetically

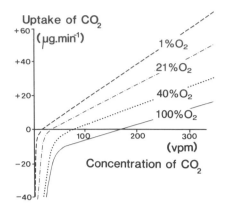

Fig. 6.1 Influence of oxygen upon uptake of carbon dioxide by plants at various concentrations of carbon dioxide. Positive assimilations represent net fixation of carbon dioxide into sugars and carbohydrates whilst negative balances represent higher rates of respiration and photo-respiration. The intercepts on the horizontal axis (where imports of carbon dioxide equal exports of carbon dioxide) are known as the 'compensation points'.

fixed carbon dioxide and there have been attempts to develop crop plants with very low levels of photorespiration which thereby may be more efficient.

The 'problem' of photorespiration really hinges on the fact that the active site of the carbon dioxide trapping enzyme RubisCO (see Ch. 2 and Appendix, Sect. 8.3) in chloroplasts will permit either carboxylation or oxygenation. Subsequent evolution of carbon dioxide by photorespiration comes about after further metabolism of the oxgenated products. Recent evidence has suggested that, instead of being a completely wasteful process (in terms of losses of energy devoted to capture of carbon dioxide), photorespiration may also be a vital protective device to permit the dissipation of light energy when amounts of carbon dioxide are limiting, and low, relative to oxygen. Such conditions would favour the production of harmful free radicals such as superoxide ($O_2^{\cdot-}$) but, by recycling carbon compounds, they would also enable the retention of a higher internal carbon dioxide concentration so that these effects would be minimised. It would appear that the oxygenase reaction of RubisCO is an inevitable consequence of the carboxylase reaction mechanism whereby an essential enzyme-bound carbon substrate (ribulose-1,5-*bis*-phosphate, see Appendix, Sect. 8.3) isomerises to an intermediate which just happens to be sensitive to both oxygen attack or carboxylation and therefore the combination has had to be retained by natural selection. This would account for the ubiquitous occurrence of both carboxylase and oxygenase activities together during carbon dioxide fixation throughout the plant kingdom, even in C_4 species (see later) where photorespiration is suppressed because the internal concentrations of carbon dioxide are so high.

(d) Hyperoxia

Potential toxic effects of raised level of oxygen upon humans have been much studied. Concentrations greater than ambient have been used medically for some time to treat certain cancers, gangrene, multiple sclerosis and lung ailments. Excess oxygen is also involved in human activities such as diving, climbing, flying and space travel. This has lead to a large number of animal model experiments to evaluate, if possible, the short- and long-term consequences of elevated oxygen concentrations at high or low atmospheric pressures. Extrapolation to predict likely human effects is fraught with the usual difficulties experienced with studies of animals exposed to experimental fumigations with atmospheric pollutants but these have certainly shown that the damaging effects of oxygen vary with species, age, organ, tissue and diet, which adds to the problems of interpretation. The common features that do emerge appear to show that mitochondrial disturbance or damage may be detected in the tissues of organs that work hard (e.g. hearts and livers) and that tissues producing fresh cells may be slowed down such as those in bone marrows which produce red blood cells, or in the testes which form sperm. There is also general agreement that the damaging effects of oxygen may be attributed to enhanced free-radical formation, the significance of which has been covered elsewhere (especially Ch. 5 and Appendix, Sect. 8.2) and more fully by Halliwell and Gutteridge (see General Bibliography, at the end of this book).

6.2 CARBON DIOXIDE

(a) The 'greenhouse effect'

Levels of carbon dioxide in the lower atmosphere (troposphere) are rising year by year (Fig. 6.2) but, in terms of the whole period of the evolution of the Earth, present concentrations of carbon dioxide and their changes over the past century are very low indeed. The amount of carbon dioxide in the troposphere now and in the past has been controlled by exchanges between the atmosphere and the biosphere by a process known collectively as the 'carbon cycle' (Fig. 6.3). As mentioned earlier (Sect. 6.1a), photosynthesis pulled the levels of

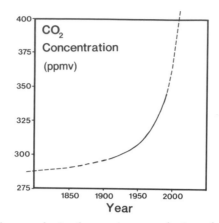

Fig. 6.2 Global atmospheric changes in atmospheric carbon dioxide with time. The solid part of the graph represents actual measured values and the dashed regions are those most likely to happen in the future or to have occurred in the past.

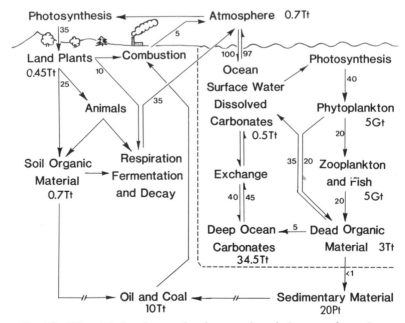

Fig. 6.3 The global carbon cycle where total pool sizes are shown in large type and fluxes in small type. The units of the latter are Gt per annum (where t is 1000 kg, G, giga, is 10^9, T, tera, is 10^{12}, and P, peta, is 10^{15}).

carbon dioxide from very high levels (> 75 %) 2 billion years ago down to below 200 ppmv during the last Ice Age. Sampling and analysis of trapped gases in ice from the polar regions has shown that levels rose slowly to about 270 ppmv before the Industrial Revolution. Increased burning of fossil fuels has since increased these 'low' levels by one-fifth to 350 ppmv over the last hundred years (1886–1986).

Future problems stem from the exponential nature of the upwards drift of atmospheric concentrations (Fig. 6.2). Over the period 1860–1960, the rise has been only 9 % but by the end of the millennium the increase over the century as a whole is expected to have been 25 % and levels of carbon dioxide may continue to rise faster thereafter. These increases are without doubt due to human activities. By burning large quantities of fossil coal and oil, which still represent only a fraction of carbon dioxide trapped by plants long perished, or by replacing tropical rain forests of large biomass with grasslands and waste scrub, the net balance of the natural carbon cycle has been tipped in favour of much more carbon dioxide being returned to the atmosphere than 100 years ago. Volcanic activity, natural fires and other phenomena may have caused a slight increase in the return of more carbon dioxide than is normally provided by steady decay processes associated with soils or seas (Fig. 6.3) but these natural processes have occurred repeatedly, before and since the last Ice Age. Within the last millenium, however, it is quite clear that it is increasing industrial activity and rapid population growth that have caused the levels of carbon dioxide to rise so significantly.

When expressed in global quantities of carbon exchanged over 1 year the full extent of this imbalance may be fully appreciated. The 5 Tt (where T, tera, is 10^{12}; t, tonnes, is 1000 kg) of fossil fuels burned in 1976, for example, contributed an extra 2.3 ppmv to the atmospheric carbon dioxide levels and of this yearly contribution, 1.6 ppmv was either absorbed by the oceans or caused the further enhancement of global photosynthesis (see later). This left a net increase of 0.7 ppmv carbon dioxide in the atmosphere by the end of 1976.

At ground level, there are considerable variations in temperature as we are all too well aware. There are also quite different temperature gradients upwards through the different atmospheric zones (Fig. 1.4). From a consideration of pollution and plant growth, the average decrease of 1 °C for every 300 m rise in height upwards through the troposphere is the most significant. On a global scale, local variations of temperature make very little difference; normally the mean global temperature never varies within more than a degree either side of 12 °C. If this pattern should change, even by a degree, then the consequences are likely to be considerable. For example, between 6000 and 8000 years ago, a temperature rise of 2 °C was responsible for melting the ice sheets spread across both hemispheres and brought about the end of the last Ice Age.

Atmospheric carbon dioxide does not absorb incoming short wavelength, ultraviolet solar radiation to any great extent but is capable of strongly absorbing long wavelength infrared radiation which is otherwise re-emitted from the surface of the Earth outwards through the atmosphere into space. This means that any increase in levels of atmospheric carbon dioxide will prevent heat being lost from the planet and cause the mean global temperature to rise. This 'warm-blanket' possibility has come to be known as the 'greenhouse effect'. As a consequence of this man-induced change, experts warn that more polar ice will melt and the sea-level rise to cause extensive flooding of low-lying areas in the world. The full extent of global flooding, they claim, has not yet been experienced because since 1932 man has been building water reservoirs fast enough to cause a 26-year delay which has mitigated rising sea-levels equivalent to 32.5 mm less than the rise should have been. Even so, the rise in sea-level since 1932 has been 64 mm and, assuming water reservoirs are still created at the same rate, the sea-level will be 85 mm higher by the year 2000 than in 1932. Thereafter, the predicted rate of increase is so great (and the rates of reservoir creation would make little difference) that rises of 2 m or more are highly likely. At the same time, other areas of the world would become more arid or prone to erosion and lead to an increase in the areas of deserts. The raising mean temperatures, for example, from 12 °C to 14 °C, would undoubtedly enhance net photosynthesis by plants, especially those of the C_4 type (see Sect. 6.2c). This would also have the effect of removing proportionally more of the extra carbon dioxide produced by the extra combustion. Unfortunately, this would be so small a fraction of the gross amount that it would be of little benefit in alleviating the problems of flooding.

Establishing what has actually happened to the atmosphere in the past and predicting exactly what might happen in the future is far from easy. Over the period 1900–1940, mean global air temperatures rose in line with what would have been predicted by the 'greenhouse effect' due to rising levels of atmospheric carbon dioxide. Yet in the period 1940–1970, the average yearly global air temperature actually fell slightly despite the fact the rate of increase of atmospheric carbon dioxide was faster. It has been suggested that other natural processes could have been involved. Increasing the amount of dust in the atmosphere due to volcanic activity could have caused temporary cooling or, alternatively, we may have entered a natural cold cycle which was not quite as cold as it should have been because of the 'greenhouse effect'. Once we left this phase then the full consequences of extra atmospheric carbon dioxide might again be expected and there is every evidence that this has been the case since 1970.

The basic causes for the slow rise in carbon dioxide concentrations from the last Ice Age to the Industrial Revolution and the reasons why the 'greenhouse effect' did not cause an immediate warming effect are probably linked to oceanic processes. Levels of atmospheric carbon dioxide are largely regulated by its partial pressure in the surface waters of the oceans where photosynthesis by the phyto-plankton also takes place. The abundance of ^{13}C in global atmospheric carbon dioxide at present is 1.11 % as compared to that of the major isotope (^{12}C) whilst ^{14}C levels are less than 10^{-10} %. In fossil fuels, on the other hand, all the ^{14}C has decayed and ^{13}C levels are lower than in the present atmosphere. During photosynthesis, there is a discrimination between the two natural isotopes of carbon whereby $^{12}CO_2$ is preferred to $^{13}CO_2$ by the carbon dioxide-fixation enzyme, RubisCO. By comparing the $^{13}C/^{12}C$ ratios in planktonic Foraminifera from surface waters of the oceans with benthonic Foraminifera which occur in deep waters, it is possible to measure the fraction of carbon dioxide returning from upwelling deep waters that is then photosynthesised. It is also possible to measure the isotopic exchange of carbon along the flow paths of deep waters and the amounts of carbon falling downwards from the organic debris of the surface waters. By deep ocean core sampling and separate analysis of planktonic and benthonic Foraminifera, it is therefore possible to take the record back over time well beyond that achieved by polar ice core gas sampling. The record of likely carbon dioxide concentrations in deep oceans and in surface waters in contact with former atmospheres now goes back over 350 000 years. By comparing the patterns of change with known periods of polar ice retreat and advance, it has been clearly shown that rises in atmospheric carbon dioxide precede climatic warming and vice versa. So the complacency built up during 1940–1970, because warming did not occur and therefore the 'greenhouse effect' could be ignored, was not justified. In the next century there is now no doubt that the consequences of our present fossil fuel burning will have severe climatic and geographical implications which should not be overlooked in any decisions concerning fossil fuel burning versus nuclear power generation.

As Sect. 5.1(b) has already mentioned, carbon dioxide is not the only gas in the atmosphere to absorb reflected long wavelength radiation coming back from the surface of the Earth. Methane, ozone, nitrous oxide and halocarbons (or CFCs, e.g. aerosol propellants, etc.) also have similar absorbance properties to carbon dioxide. Moreover, on a molecular basis, some of these trace halocarbon gases trap heat as much as 10 000 times as effectively as carbon dioxide molecules. Thus a small rise in such compounds will produce a disproportionate absorption of reflected radiation. The likely releases of halocarbons in the future are difficult to predict but they are probably in the range of from 2.5 to 5.1 % per annum (as a percentage increase on top of the atmospheric concentrations in 1985) up to year 2025, and thereafter, increases of 2.1 % per annum are possible. Such rates of release could therefore take levels of halocarbons from 0.6 ppbv in 1985 to 6–8 ppbv in year 2050. In turn this may result in a disproportionate increase in the amount of globally trapped thermal energy, amounting to as much as 70 % of that caused by carbon dioxide alone because halocarbons are more efficient absorbers than carbon dioxide.

To compound the problem, another important long-wavelength absorber like nitrous oxide (currently around 300 ppbv in the Northern Hemisphere) may also rise to 350 ppbv by year 2050 because of increased microbial denitrification and excessive use of artificial fertilisers (see Sect. 3.1b). Methane (currently 1.5 ppmv) may similarly contribute to the 'greenhouse effect' because it is increasing at the rate of 1 % per annum to 2.5 ppmv by year 2050.

The scenario for the future is therefore much worse than predictions of global increases in carbon dioxide alone would suggest. Taken together, the increases in atmospheric levels of halocarbons, nitrous oxide, methane and tropospheric ozone could double the amount of global thermal heating predicted for increases of atmospheric levels of carbon dioxide alone. The climatic, geographical (i.e. flooding), agricultural and economic consequence of this is undoubtedly one of the most serious problems that the present generation now being born, and those yet to come, will have to face.

(b) CO_2 enrichment and deprivation in crops

During commercial horticulture, growers often deliberately increase the atmospheric levels of carbon dioxide inside their glasshouses. Originally, oil-fired burners were used to enrich glasshouses with carbon dioxide but, as the costs for heating soared, the practice of redirecting flue gases directly into the glasshouses from heating systems (which burn either propane or natural gas) became more common. The possible benefits in growth that may be obtained from such carbon dioxide enrichment are shown by Fig. 6.4. Although there are variations even between cultivars of the same crop species, as a general rule an additional 30 % of growth (including fruit production) may be obtained by raising the glasshouse atmosphere to 1000 ppmv carbon dioxide; thereafter, the gains are not significant. The pollution problems that occur with these procedures (mainly from oxides of nitrogen) have been covered elsewhere (Sect. 3.2d). Incomplete combustion, which causes release of ethylene or oxides of

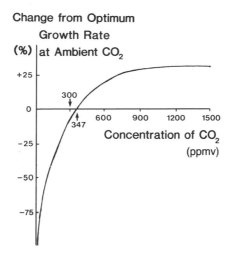

Fig. 6.4 The likely advantages of enriching atmospheres above horticultural crops with carbon dioxide. Note that above 1000 ppmv little further improvement is to be gained and that a heavy penalty is paid if levels of carbon dioxide fall below ambient.

nitrogen pollution, may be sufficient to abolish all the possible gains by certain crops if attention is not paid to adequate control of burners, flow of flue gases and ventilation rates. Unfortunately, as the ventilators of a commercial glasshouse are opened during the day for cooling, concentrations of carbon dioxide quickly fall from 1000 ppmv to ambient (350 ppmv) and thus the plants lose any gain derived from enrichment. In fact, it is likely to be worse than that because the crop in the glasshouse will be fixing carbon dioxide so rapidly under warm conditions that the glasshouse atmospheric concentration of carbon dioxide may fall to below 100 ppmv which would cause over 60 % less growth than in an equivalent crop grown outside the glasshouse. Consequently, control systems that monitor and quickly correct carbon dioxide concentrations inside glasshouses under open or closed ventilation conditions are vital if growers are to achieve their objectives – bigger and better quality fruits and flowers, faster, at least cost, at the right time for the market.

Significant reductions in growth due to deprivation of carbon dioxide are frequently experienced by plants. In crops growing close together, such as in tropical and subtropical forest canopies, levels of carbon dioxide may be reduced well below ambient. Over long periods of time, competition for these reduced amounts of carbon dioxide between species within canopies favours those that are better able to capture carbon dioxide than their neighbours. In subtropical climates, this evolutionary pressure has caused the emergence of plants with more efficient carbon dioxide-trapping and concentrating mechanisms.

(c) C_4 photosynthesis and CAM metabolism

All plants trap atmospheric carbon dioxides by means of the enzyme RubisCO with the assistance of other enzymes of the Calvin (C_3) cycle (see Appendix, Sect. 8.3). Such a process appears to have existed since the evolution of photosynthesis (about 2.5 billion years ago) and all flowering plants, arising about 150 million years ago, have retained this mechanism. On a number of occasions during the natural selection of plants, however, a significant modification of conventional C_3 photosynthesis involving four carbon (C_4) acids has occurred and been retained in a wide range of plant species which include some very important crops such as maize, sorghum, sugar-cane and millet. Modern understanding of this modification would be now to regard C_4 photosynthesis, not as a replacement of C_3 photosynthesis, but as a beneficial adjunct to this basic process under certain environmental conditions. Perhaps the best way to appreciate the importance of this modification to photosynthesis is to compare it to the improvement to the performance of a combustion engine one would gain by adding a turbocharger.

All three types of C_4 plants discovered so far have different arrangement of leaf cells as compared to normal C_3 plants. Outer mesophyll cells completely surround inner bundle sheath cells which, in turn, enclose the vascular elements.

The basic metabolic feature of C_4 photosynthesis is the primary assimilation of carbon dioxide by the carboxylation of PEP or phosphoenolpyruvate (see Appendix, Sect. 8.3) catalysed by PEP carboxylase (a more efficient carbon dioxide-trapping enzyme than RubisCO, which is found in the cytoplasm of the mesophyll cells) to form oxaloacetate (see Fig. 6.5). This C_4 compound may be then reduced to malate (or, in some species, transaminated to aspartate) each of which may be then transported from the mesophyll cells to the bundle sheath cells through cell-to-cell connecting channels (plasmodesmata). Intermediate metabolic steps in different C_4 variants may differ but the carbon dioxide released by decarboxylation is then refixed by RubisCO in the plastids of the bundle sheath cells. The pyruvate (or alanine) left over (see Fig. 6.5) is transferred back into the mesophyll cells where it is converted back into precursor PEP by catalysis with pyruvate phosphodikinase – thus completing the cycle of C_4 photosynthesis across the two types of cell.

The essence of the C_4 process is that advantage of the higher affinity (or lower K_m) for carbon dioxide of the enzyme PEP carboxy-

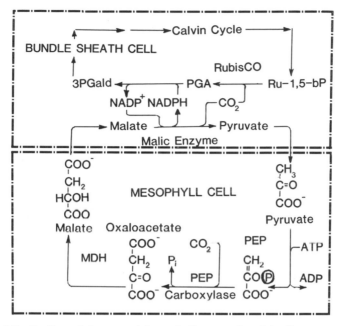

Fig. 6.5 Outline of the essential metabolic steps found in C$_4$-type photosynthesis. Three different types of metabolism are known but those shown here are characteristic of maize, sugar-cane and sorghum. Abbreviations: PEP, phosphoenolpyruvate; MDH, malate dehydrogenase; PGA, phosphoglycerate; 3PGald, 3-phosphoglyceraldehyde; Ru-1,5-bP, ribulose-1,5-*bis*-phosphate.

lase over that of RubisCO. Furthermore, the oxygenase activity of RubisCO is suppressed by the carbon dioxide-pumping action of PEP carboxylase thereby raising the effective concentration of carbon dioxide in the bundle sheath plastids and favouring the carboxylation activity of RubisCO. Any photorespiratory carbon dioxide that escapes from the bundle sheaths is then immediately recaptured by PEP carboxylase. As a consequence, C$_4$ plants show very low internal concentrations of carbon dioxide but are still much more effective by comparison with C$_3$ plants in being able to reduce carbon dioxide concentrations around a leaf. Within stands of vegetation in warm climates where amounts of carbon dioxide are limiting, this property is of considerable evolutionary advantage. Moreover, most (but not all) C$_4$ plants, unlike C$_3$ plants, do not show saturation of photosynthesis at higher light fluences because the oxygenase activity of RubisCO is still inhibited, so higher overall rates are ultimately possible as well.

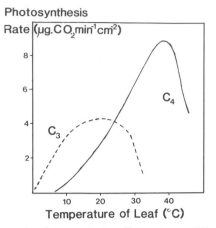

Photosynthesis
Rate $(\mu g.CO_2 min^{-1} cm^2)$

Temperature of Leaf (°C)

Fig. 6.6 Differences in photosynthetic efficiencies at different temperatures of normal C_3-type plants and those having the additional C_4-type of metabolism which concentrates carbon dioxide in the tissues. Below 20 °C, C_3 plants are favoured whilst above 25 °C, C_4-type plants are more effective.

In the majority of C_3 species, the rates of photorespiration (see Sect. 6.1c) increase more than the rates of photosynthesis as the temperature rises. Consequently, C_3 plants show a temperature optimum between 10 and 28 °C (Fig. 6.6). With the additional inhibition of the oxygenase function in C_4 species, the temperature optimum of C_4 plants is raised to 40 °C. Therefore C_3 plants tend to predominate in cool moist environments but C_4 plants are to be found in warm or hot environments especially in close proximity to other plants. In terms of any 'greenhouse effect', the consequences of raising of mean global temperatures will be to favour plants with C_4 photosynthesis over those confined to using just C_3 mechanisms. Theoretically, areas of the Earth which become warmer may show more net photosynthesis and biomass accumulation by the existing vegetation. Unfortunately, this is unlikely to be the case in practice because water evaporation rates from land surfaces will also increase and cause larger areas to become more arid. Some plants have already adapted to water deprivation but this limiting factor to growth could become more important as the full global consequence of the 'greenhouse effect' materialises.

A specialised group of plants show yet another type of photosynthesis (again based on C_4-acids) called Crassulacean acid metabolism (CAM). There are many metabolic similarities between CAM and C_4 plants. The principal differences between them are that, malic or aspartic acids are decarboxylated almost immediately in C_4 plants, whilst in CAM plants they are accumulated during darkness and decarboxylated later the next day. In other words, there is a temporal rather than a spatial separation of the main carbon dioxide-fixing activities of PEP carboxylase and RubisCO. As a consequence, CAM plants lack the double cell cooperation of C_4 plants but have succulent mesophyll tissues instead.

Because stomata are normally open at night and closed during the day within the reversed diurnal cycle of CAM photosynthesis, the water-use efficiency (i.e. $g\ H_2O$ transpired $g^{-1}\ CO_2$ assimilated) of CAM plants is very high as compared with those engaged in C_4 photosynthesis. This in turn is about twice as efficient in terms of usage of water as conventional C_3 metabolism because transpiration rates are reduced at the lower temperatures and higher humidities of night-time in CAM plants and the intercellular levels of carbon dioxide are lower in C_4 plants by comparison to those in C_3 plants. Consequently, CAM photosynthesis is usually associated with plants that grow in hot, dry habitats with unpredictable rainfall.

When the full consequences of the 'greenhouse effect' are expressed in the next century, CAM plants will be the best survivors in those land areas that become drier, whilst the phytoplankton will flourish in the oceans and be responsible for an even more significant proportion of the global carbon budget (see Fig. 6.2). This is because the surface waters of the oceans will be warmer and therefore contain more dissolved carbon dioxide and the global wet surface area will be greater as more coastal regions are flooded. This, in turn, assumes that the effectiveness of the ozone layer in the stratosphere to absorb the ultraviolet element of incoming solar radiation is not simultaneously impaired. Should this also happen then the surface plankton could be the most affected component of the biosphere and that would have severe implications for the productivity of ocean fisheries.

(d) Asphyxiation

The fact that animals use carbon dioxide in a manner different to plants has been known ever since Joseph Priestley did his experiments on combustion and respiration between 1771 and 1772 (see Appendix, Sect. 8.6). When he burned a candle in an enclosed space, the air inside would no longer support burning, and mice died in the residual air but plants lived on for several weeks. After this period, a candle would burn or mice could breathe the air that was left. He correctly concluded that plants used and changed air in a manner different from animals. Subsequently, he followed up these observations, co-discovered nitrogen and laid the foundations for our understanding of respiration (see Appendix, Sect. 8.6).

The presence of other gases in confined spaces and lack of the ability to breathe under certain conditions has been recognised by miners long before Priestley did his experiments. 'Black damp', as they called it, consists of air with more than 13 % carbon dioxide and was tested for in a variety of ways. Lighted candles or tapers are extinguished at concentrations of less than 17 % oxygen, or when the carbon dioxide rises to 10 %, but this had unfortunate repercussions for the early miners if methane and other explosive gases were present. Developments of safety lamps improved this situation but, if acetylene was used, they found that the flame was not extinguished until levels of 29 % of carbon dioxide had been achieved. A much better safeguard they found was to take mice along and lower them into any suspect shaft.

Nowadays, carbon dioxide is widely used in industry and likely to be experienced in relatively high concentration by those who work in a number of environments (not just mines) unless precautions are taken. This is especially important in those processes which use carbon dioxide as a refrigerant, as a weak acid for the manufacture of drugs or white lead, in water treatment, whilst filling fire extinguishers, or during food preservation, welding, purging pipelines and carbonated drink production. Sealed systems such as submarines or spacecraft involve perhaps the greatest exposure potential for humans to carbon dioxide but controls are also necessary in industries which involve fermentation (e.g. brewing and wine making). During sugar-beet processing, for example, the juice is treated with lime and the calcium saccharide is precipitated by carbon dioxide. All these operations and industries have produced their fatal accidents from time to time. The most bizarre was the death of four Irish salvage workers who tried to remove a cargo of rotting apples from the wrecked liner *Celtic* so that it could be floated off the rocks outside Cork harbour.

All investigations have shown that it is the deprivation of oxygen that is the cause of harm. Breathing is very deep and rapid at 12 % oxygen and severe headaches are experienced. Unconsciousness sets in below this level of oxygen and death occurs when levels of oxygen are between 8 % and 5 %. The physiological effect of elevated carbon dioxide is to stimulate the respiratory centre of the brain near the top of the spinal cord. Five per cent carbon dioxide in air produces strong stimulation and is used medically for the resuscitation of gas victims or inducing respiration in new-born babies. Studies in submarines have shown that atmospheres containing 20 % carbon dioxide may be tolerated for long periods with little effect provided that oxygen levels are adequate. At 30 % carbon dioxide, blood pressure increases and hearing becomes difficult whilst 30 min at 50 % produces signs of intoxication. Unconsciousness is produced by 70 % carbon dioxide within a very short period. The recommended Threshold Limit Value or Occupational Exposure Limit of 5 % (5000 ppmv) carbon dioxide for 8 hours daily would appear to have a wide margin of safety.

6.3 CARBON MONOXIDE

(a) Sources and sinks

If we accept that carbon dioxide is not generally a pollutant with respect to animals or plants directly but to the stability of the climate of the Earth, carbon monoxide is the next most widely distributed and abundant atmospheric pollutant in the troposphere. Emissions of carbon monoxide caused by man exceed those of all other pollutants combined and in 1974 amounted globally to 0.6 Gt (where G, giga, is thousand million, 10^9; t, tonnes, is 1000 kg) per annum – an enormous rate and still rising. These emissions are concentrated in the Northern Hemisphere and show distinctive rhythms – daily, weekly and seasonal – depending upon human activity. Levels of atmospheric carbon monoxide are highest during winter in the Northern Hemisphere due to space heating and lowest during summer months, with the differences between the seasons becoming more marked as time passes. Above heavy traffic, carbon monoxide levels are often well above 50 ppmv and may even exceed 100 ppmv for long periods in tunnels, garages and loading bays. Natural sources also produce significant quantities of carbon monoxide. About 5 % of that produced by the activities of man arises from the surface of the oceans and another 5 % arises from a variety of natural processes such as volcanoes, electrical storms, 'marsh gas' from rotting vegetation and a variety of other biological sources.

In areas away from major sources, the atmospheric levels of carbon monoxide are relatively constant which means that significant natural sinks exist to remove this vast excess of carbon monoxide. The problem has been to identify the major natural sinks and their relative importances and it would appear that five possibilities exist, namely, the oxidation to carbon dioxide in the troposphere by hydroxyl radicals, oxidation in the stratosphere or absorption by soil microbes, the oceans, or plants and animals.

Carbon monoxide is produced by incomplete combustion and is rapidly oxidised to carbon dioxide in the atmosphere by hydroxyl radicals (Reaction 6.1, see also Fig. 5.2) and may add to any 'greenhouse effect' carbon dioxide may have upon climate (see Sect. 6.2a). Other reactions with, molecular oxygen or oxides of nitrogen, which have higher activation energies, are less significant. The rapidity of removal of carbon monoxide therefore depends to a large extent on the presence of sufficient hydroxyl radicals ($OH^{\bullet}$) which, in turn, requires atomic oxygen and water (Reaction 5.7). Consequently, hot, bright and wet atmospheric conditions are most favourable for hydroxyl production and carbon monoxide removal – dull, overcast, cold but dry winter days are far from ideal. Any carbon monoxide that rises above the troposphere into the stratosphere will be more prone to oxidation to carbon dioxide by atomic oxygen (Reaction 6.2) or ozone but this is counterbalanced by photolysis of the heavier gas (Reaction 6.3).

$$OH^{\bullet} + CO \longrightarrow CO_2 + H \tag{6.1}$$

$$CO + O + M \longrightarrow CO_2 + M \tag{6.2}$$

$$CO_2 + light \longrightarrow CO + O \tag{6.3}$$

Numerous biological mechanisms have been proposed for the removal of carbon monoxide but the soil appears to have a variety of soil microbes which will accept carbon monoxide and either oxidise it to carbon dioxide or reduce it to methane. For example, the anaerobic methane producers *Methanosarcina barkerii* and *Methanobacterium formicum* may generate hydrogen (Reaction 6.4) which may then be used to reduce carbon monoxide or carbon dioxide (Reactions 6.5 and 6.6). Whilst attempting to understand the relative importance of possible removal mechanisms of carbon monoxide by soil bacteria, it is relatively easy to show that various microbes utilise carbon monoxide but very difficult to extrapolate such local microcosms on to a wider global scale. Closed systems with soil and atmospheres containing carbon monoxide (120 ppmv) show complete absorption of the carbon monoxide into the soil within 3 hours. Sterilising the soil, treating with antibiotics, salting or removing oxygen from the system prevent soil uptake of carbon monoxide and demonstrate that this is an aerobic microbial process. Different soils remove carbon monoxide at different rates but cultivated soils are much less efficient in doing this than natural soils high in organic matter and acidity. Agricultural processes thus appear to select against those microbes most efficient at carbon monoxide removal. The health of forest soils is therefore vital not just for protection against acid deposition (see Ch. 4) but also to renovate our atmospheres by reducing carbon monoxide levels.

$$CO + H_2O \longrightarrow CO_2 + H_2 \tag{6.4}$$

$$3H_2 + CO \longrightarrow CH_4 + H_2O \tag{6.5}$$

$$4H_2 + CO_2 \longrightarrow CH_4 + 2H_2O \tag{6.6}$$

Some global estimates of the total potential capacity for soil microbial uptake of carbon monoxide have been put as high as 14 Gt (G, giga, is 10^9; t, tonnes, is 1000 kg) per annum; which is well in excess of what is actually being discharged by human or oceanic activities. By contrast, other calculations have estimated the global rate of removal to be 0.45 Gt per annum which is less than the annual rate of production (0.6 Gt per annum). Large assumptions have been made in both estimates because of uncertainties over soil distribution which accounts for the high discrepancy. It is certain that soil microbial uptake is the major global mechanism for carbon monoxide removal but it is a slow temperature-dependent process. This would explain the rhythmical nature of the changes in levels of global atmospheric carbon monoxide across the seasons. Only when the northern forest soils have warmed sufficiently can they supplement the steady uptake of carbon monoxide by tropical forest soils. It also follows that if more tropical and temperate areas are brought into cultivation, the cyclical seasonal changes in global carbon monoxide concentrations will become more pronounced as current trends are indicating. Should northern forest soils also suffer damage by other types of pollution (Chs. 4 and 7) then the overall levels of carbon monoxide, and thereby carbon dioxide (Reaction 6.1), may rise and enhance the 'greenhouse effect'.

The oceans are the major sink for carbon dioxide (Sect. 6.2) but not for carbon monoxide. A function of partial pressure, the solubility of carbon monoxide in seawater is much less than that of carbon dioxide or sulphur dioxide. However, oceanographic investigations have shown that surface waters of the oceans may contain from seven to ninety times the theoretical carbon monoxide concentrations that would be expected from gaseous uptake from the atmosphere. In other words, marine biological activity contributes substantial quantities of carbon monoxide to surface waters and, by exchange with the atmosphere, is capable of providing even more carbon monoxide. The organisms that appear to be partially responsible for such carbon monoxide formation are the marine algae. Certain siphonophores (primitive carnivorous animals), which prey upon this plankton, are interesting because they have gas-filled cavities called 'pneumatophores' which they use to control flotation so that they move with distinctive daily rhythms from deeper waters upwards during the evening and downwards during the morning. The gas in the pneumatophores is mainly carbon monoxide which means that, at great depth (450 m), the partial pressure of carbon monoxide is equivalent to 45 atm and the biochemistry of their cytochrome oxidase and other processes must be very unusual to allow them to survive such pressures.

(b) Plant uptake

Oxidation of carbon monoxide to carbon dioxide by plants has often been detected. Using ^{14}C-labelled carbon monoxide, it has been shown that the radioactivity appears in the amino-acids serine and glycine. If respiration is inhibited with dinitrophenol then a little carbon monoxide still appears to be fixed by a photosynthetic process although the major oxidation of carbon monoxide is linked to the cytochrome oxidase activity of their mitochondria (see Appendix, Sect. 8.4).

Comparisons of measured rates of carbon monoxide exchange with those of carbon dioxide fixation by plants under various atmospheric conditions have produced wide differences. Some plants, especially trees, do not take up carbon monoxide at all whilst beans have quite high rates of uptake. But others, especially green algae, have their nitrogen metabolism inhibited by carbon monoxide and there are even some seaweeds that are capable of emitting carbon monoxide. Any judgement on the relative significance of vegetation to that of the soil biomass as a mechanism for carbon monoxide removal is therefore very difficult to make. Perhaps an uptake of carbon monoxide by vegetation one-quarter that of soil might be a fair global estimate. More realistic studies of carbon monoxide uptake by both plant and soils are urgently needed in order to evaluate the capacity of vegetation to reduce high local carbon monoxide levels due to heavy traffic in urban areas.

(c) The oldest industrial poison

Ever since man learned how to break stones by heat or to burn wood with restricted flow of air to produce charcoal he has been affected by carbon monoxide. Whilst smoking is the single, most important, source of carbon monoxide for humans, most of the hazards at work connected with the gas come from motor vehicles, especially during their repair, testing, racing or servicing. Blasting, fire fighting, methanol production, wood distillation and even cooking over charcoal in French restaurants (an affliction known to French doctors as *'folie des cusiniers'*) have similar hazards. If combustion processes are improperly ventilated (e.g. in smelting processes or in home and office heating systems) similar problems could arise. Any burner which has a surface cooler than the ignition temperature of the gas phase of the flame will start to form carbon monoxide. In water heaters, where coils containing water cannot rise above 100 °C, large quantities of carbon monoxide are often given off and, frequently, are not ventilated away properly. This may become a bigger problem in the home, especially the kitchen, as draughts are restricted to curtail the costs of space heating.

Exposure of humans to carbon monoxide may cause a variety of clinical effects (Table 6.1). Generally, there is a fair relationship

Table 6.1   Effects of carbon monoxide and their likely accompanying carboxyhaemoglobin blood (COHb) content in humans

Exposure range[†] (ppmv)	Effect and symptoms	COHb[‡] blood conc. (%)
0–10	No discomfort or effect	0–2
10–50	General tiredness, impaired vigilance and reduction in manual dexterity	2–10
50–100	Slight headache, pronounced tiredness and irritability	10–20
100–200	Mild headache	20–30
200–400	Severe headache, visual impairment, nausea, general weakness and vomiting	30–40
400–600	As above, but with greater possibility of collapse	40–50
600–800	Fainting, increased pulse rate and convulsions	50–60
800–1600	Coma, weak pulse and possibility of death	60–70
above 1600	Death within a short period	70 plus

† After 2 hours exposure.
‡ Likely equilibrium carboxyhaemoglobin/oxyhaemoglobin ratios in the blood after some time whilst resting (although they may be achieved three times as fast during heavy work).

between symptom and the level of the gas but responses are hastened by heat, humidity and exertion. At work, the young are usually more susceptible than the older workers but alcoholism, obesity, old age, heart conditions and lung diseases all exacerbate the problem. Following repeated exposure some degree of compensation occurs such as, for example, the production of more red blood cells. This type of conditioning has been found in workers in the steel industry, as well as smokers, who appear to be able to tolerate slightly higher levels before the appropriate symptoms develop. Without this type of adaptation there are clinical differences between oxygen deprivation (hypoxia or anoxia) and carbon monoxide poisoning. In the former, respiratory symptoms precede nervous symptoms but, in the latter, this order is reversed. This has led to the suggestion that the driving out of oxygen from red blood cells (see below) is not the only physiological effect of carbon monoxide. Apart from anoxia (lack of oxygen), which impairs nerve and brain cells, there may be another type of toxic effect in all body tissues. This has not yet been fully evaluated but it is thought that levels as low as 3–5 % saturation of blood with carbon monoxide may elicit these subtle effects within the central nervous system.

The progressive symptoms listed in Table 6.1 may approach without warning. Carbon monoxide is non-irritating, tasteless and without smell. Only the slightest presence of other contaminants such as hydrocarbons, mercaptans, ammonia or traces of other materials give an advanced warning of any hazard within fumes. It was a long time before carbon monoxide poisoning became to be universally accepted as a definite disease in the industrial and urban environment because many of the early symptoms caused by carbon monoxide are also associated with repetitive factory work or stressful situations such as driving. The question: 'Does enhanced blood carbon monoxide levels lead to more accidents at work, on the road or in the home?' still remains unanswered because so many other factors are also involved. On balance, atmospheric levels of carbon monoxide must be taken as a significant contributory cause towards misjudgements prior to accidents.

Related to carbon monoxide poisoning is the disease, Shinshu myocardosis in which individuals exposed to high levels of carbon monoxide over long periods have defective heart valves as well as angina pectoris and arteriosclerosis. It was first detected in 1955 in the village of Kinasa, Japan where work on silkworm production continued during cold winters in tightly sealed, heated rooms.

(d)   Blood biochemistry and pollution

Haemoglobin, the red pigment of red blood cells, consists of the protein globin and a tetrapyrrole ring system called 'haem' which contains ferrous iron. If the iron should become oxidised and form a complex known as 'methaemoglobin', as has already been suggested during exposure to oxides of nitrogen (Sect. 3.3), then it is unable to bind oxygen. Oxidation of haemoglobin to methaemoglobin occurs naturally as red blood cells become older and this type of ageing is accelerated by the additional presence of hydrogen peroxide. Mechanisms of electron scavenging, involving glutathione peroxidase and catalase (Ch. 5) normally ensure that this does not occur prematurely but free-radical generation by other pollutants such as ozone or PAN also promote the early formation of methaemoglobin. Ultimately, these older blood cells high in methaemoglobin formation are more likely to be broken up by certain white blood cells (phagocytes) whilst they pass through the spleen.

During normal respiration oxygen combines with haemoglobin to form oxyhaemoglobin; an association-dissociation reaction which is dependent upon the partial pressure of oxygen, temperature and pH. Any changes of acidity in blood are largely due to alteration in blood levels of dissolved carbon dioxide and the equilibrium between different forms of carbon dioxide (Reaction 6.7) is determined by numbers of different factors. The enzyme carbonic anhydrase brings about a faster interconversion between dissolved carbon dioxide and carbonic acid whilst protons from the dissociation of carbonic acid produce a more acid environment which causes the dissociation of oxyhaemoglobin. Lactate, which is produced by contracting muscles working at low oxygen tensions, also produces acidity which then causes more oxygen to be released from oxyhaemoglobin to alleviate the local shortage. The final factor that encourages the release of oxygen from haemoglobin is 2,3-diphosphoglycerate. This substance combines reversibly with haemoglobin to change its structure in favour of oxygen release. Consequently, 2,3-diphosphoglycerate is released by tissues which are especially short of oxygen to ensure that oxygen is delivered to those tissues most in need.

$$CO_2 + H_2O \; \rightleftharpoons \; H_2CO_3 \; \rightleftharpoons \; H^+ + HCO_3^- \qquad (6.7)$$

About 7 % of carbon dioxide produced by body tissues is dissolved in the blood plasma and carried to the lungs as a dissolved gas. The bulk (70 %) is transported as bicarbonate and the remainder (23 %) reacts with haemoglobin to form carbaminohaemoglobin. At the high partial pressures of carbon dioxide in the tissue capillaries, the formation of this complex is encouraged but in the lungs where partial pressures of carbon dioxide are low it splits apart again.

Carbon monoxide also combines with haemoglobin with an affinity which is between 200 and 300 times greater than the affinity of haemoglobin for oxygen. The resultant carbon monoxide complex or carboxyhaemoglobin is extremely stable. Carbon monoxide at very low concentrations in the blood (i.e. 0.1 %, equivalent to a partial pressure of 0.5 mm of mercury) will combine with over half of the haemoglobin and immediately reduce the oxygen-carrying capacity by a similar proportion. Only very high oxygen partial pressures, such as those achieved when treating a carbon monoxide poisoning case with pure oxygen are sufficient to reverse the combination of haemoglobin with carbon monoxide.

Other oxygen-releasing mechanisms may be adversely affected by carbon monoxide. For example, levels of 2,3-diphosphoglycerate in plasma circulating through oxygen-deficient tissues are lower in the presence of carbon monoxide. The possibility that the body attempts to compensate for reductions of oxygen-carrying capacity has been considered but if it does so, understanding of the exact mechanism remains elusive. Contraction of the spleen was originally thought to take place when carboxyhaemoglobin is present in the blood but there is little evidence to support this. Increased bone marrow activity causing higher numbers of red blood cells has also been suggested but this only appears to be a short-term response towards frequent exposures of carbon monoxide.

Unfortunately, the affinity of carbon monoxide for human foetal haemoglobin is higher than that for normal haemoglobin. This means that unborn babies in the womb are especially sensitive to carbon-monoxide poisoning. The single most important factor which exposes an unborn child to greater-than-average carbon monoxide concentrations is smoking by the mother. The number of developmental and clinical defects in the new-born ascribed to deprivation of oxygen are many and mostly neurological in nature. Even when these are not fully diagnosed, the relationship between maternal smoking and low birth weight is undeniable. As with adults, it is still not possible to distinguish with any certainty the direct effects of carbon monoxide upon these babies caused by oxygen deprivation from those caused by other factors in tobacco smoke. The World Health Organisation has recently issued guidelines which recommended that levels of carbon monoxide are reduced so that the level of carboxyhaemoglobin in the blood of the general population does not exceed 3 %. This means that exposures should not exceed 50 ppmv over 30 min, 25 ppmv over 1 hour and 10 ppmv over 8–24 hours.

6.4   HYDROGEN SULPHIDE

(a)   Malodorous emissions

Hydrogen sulphide – with the notorious 'bad eggs' odour – is a highly toxic and flammable dense gas. There are few industrial uses for hydrogen sulphide (H_2S) but it is produced in large amounts as a by-product of a number of both natural and industrial processes, especially during the mining of sulphur. Over 90 % of global emissions may be accounted for by man-induced processes. Natural emissions of hydrogen sulphide, by contrast, are quite low. It occurs in certain natural gas or oil reserves and is one of the major hazards encountered during the production and refining of high-sulphur petroleum because some of the hydrogen sulphide is liberated from crude oil as soon as it reaches the surface, especially if it is heated. Decay of organic matter in sewers or of animal and plant wastes by microbial activity are additional major sources of hydrogen sulphide and other gaseous sulphides (see below). Tanneries, glue and fur-dressing factories, abattoirs, waste-treatment plants and sugar-beet processing all produce quantities of hydrogen sulphide (along with other mercaptans) and account for the fact that these are some of the least-popular neighbouring industries. Tanning is very prone to accidental discharges because the first stage of hair removal employs a paste of sodium sulphide and the second chrome-tanning process uses sulphuric acid. If care is not taken to prevent the two effluents from mixing then hydrogen sulphide is liberated. The manufacture of paper, rayon and sulphur dyes or the vulcanising of rubber are also industrial processes where hydrogen sulphide may be generated and thus special safety precautions should be taken to protect workers. The manufacture of rayon by the viscose process also produces carbon disulphide which has its own toxicological problems (TLV or OEL, 20 ppmv). In areas around pulp mills, the levels of hydrogen sulphide may be as much as 11.5 ppmv whilst even higher levels have been recorded inside such factories (up to 21.5 ppmv).

In urban areas, levels of hydrogen sulphide are normally much less than 5.4 ppbv. If they exceed this level for any length of time then substantial complaints are likely to be made by the public because people's sense of smell with respect to this gas is so acute. The threshold of human odour detection occurs between 0.15 and 1.5 ppbv whilst the rotten egg smell is clearly apparent at about three times this level. On most occasions when problems of smell occur, hydrogen sulphide is also accompanied by other malodorous substances such as carbon disulphide, CS_2, methane thiol, CH_3SH, dimethyl disulphide, $(CH_3S)_2$, and dimethyl monosulphide, $(CH_3)_2S$. The 'quality' of the odour is changed accordingly – presumably from 'bad' to 'worse'.

(b) Emission by plants and microbes

Hydrogen sulphide is slightly phytotoxic. Only at levels above 100 ppbv have necrotic lesions and tip-burn been detected on a variety of plants. Indeed it may be one of the few other instances apart from carbon monoxide where plants are less sensitive to an atmospheric pollutant than man himself. Beneficial effects upon growth have been reported with low concentrations (30 ppb) but, of more interest, is the fact that living, as well as rotting, vegetation may be a significant source of hydrogen sulphide. Young leaves appear to emit much more than old leaves. When young cucumber leaves were exposed to ^{35}S-labelled sulphur dioxide, the specific activity of the $H_2^{35}S$ emitted was over 50 % of the absorbed $^{35}SO_2$ but the amount of $H_2^{35}S$ in the form of ^{35}S-labelled sulphate was very low. This means that sulphate is probably not an intermediate in the synthesis of hydrogen sulphide from sulphur dioxide. The biological aspects of the sulphur cycle have already been covered in Ch. 2 but Fig. 6.7 illustrates how production of hydrogen sulphide from sulphur dioxide may be an alternative mechanism by which the harmful effects of

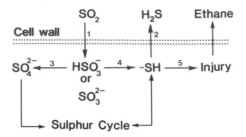

Fig. 6.7 Alternative routes of sulphur metabolism exist for atmospheric sulphur dioxide. It is thought that some SO_2-tolerant plants have either reduced uptake (1) characteristics or have the ability to metabolise more sulphite or bisulphite to reduced sulphur (4) and release more sulphur as H_2S (2) thereby reducing the rate of entry of sulphur from SO_2 into the sulphur cycle (3) and the amount of damage (5).

atmospheric sulphur dioxide upon tolerant plant tissues are mitigated.

The question: 'How much sulphur do plants in the field emit as hydrogen sulphide to the atmosphere?' still remains to be answered. Estimates have indicated that of the 100 million metric tons of sulphur dioxide and hydrogen sulphide entering the atmosphere each year by natural processes, more than 7.4 million metric tons could be liberated by plants. Why they should do this is also intriguing. It is not a means of getting rid of excess reductant because rates of sulphate reduction are less than 0.1 % of those of carbon dioxide fixation but it may be a mechanism to maintain reservoirs of specific reductants and carriers involved in sulphate reduction in order to maintain a balance between thiol or sulphydryl (-SH) and disulphide (-S-S-) groups. Levels of glutathione and cysteine involved in free-radical scavenging are carefully controlled in all biological tissues. Perhaps the release of hydrogen sulphide may be seen rather like a pressure valve whereby excess sulphur is released after sulphur metabolism within cells. During the evolution of life on Earth, one problem that must have been frequently encountered inside primitive cells would have been the storage of large quantities of sulphur in a variety of different forms. A mechanism to rid a primitive organism of this excess would have been a considerable selective advantage and may have been retained for a variety of reasons.

Micro-organisms of the soil and sewage-treatment plants also produce considerable quantities of dimethyl sulphide, $(CH_3)_2S$, dimethyl disulphide, $(CH_3S)_2$, and methyl mercaptan, CH_3SH. Recent atmospheric analyses suggest that the emissions of dimethyl sulphide by the biosphere may exceed those of hydrogen sulphide and that a large proportion of this may be coming from the oceans. Collectively, these emissions may be an important component of the biological sulphur cycle (see Ch. 2).

(c)   Accidents

Normally, the smell of hydrogen sulphide is sufficient to warn humans of the presence of this gas at levels below 20 ppbv and to cause them to retreat or to take remedial action. However, paralysis of the sense of smell occurs at higher levels (150 ppmv) so victims may be unaware of the dangers (see Table 6.2). High concentrations of hydrogen sulphide are as toxic as hydrogen cyanide (both having the same TLV or OEL of 10 ppmv) and the two are similar in effect. Respiratory failure occurs within seconds due to paralysis of the nervous control of breathing. The toxicological problems associated with hydrogen sulphide, like those of cyanide, are caused by an inhibition of electron flow through cytochrome oxidase to oxygen in the mitochondria of cells. The skin of victims of acute intoxication by hydrogen sulphide is usually grey-green in colour as are the internal organs. This is due to the formation of a sulphaemoglobin immediately after death. Such was the fate of twenty-three inhabitants of the town of Poza Rica, Mexico in 1950 when a faulty valve on a new gas field installation leaked hydrogen sulphide for only 20 min.

Most of the metabolic consequences of sublethal exposures to hydrogen sulphide stem from the partial inhibition of cytochrome oxidase with the mitochondria of all cells. This brings about an

Table 6.2   Effects of hydrogen sulphide on humans

Exposure range	Effects and symptoms
0–0.2 ppbv	No discomfort
0.2–1.2 ppbv	Odour detected
1.2–5.4 ppbv	Consciousness of 'bad egg' smell
5.4 ppbv–10 ppmv	Eye irritation threshold
10–50 ppmv	Severe eye irritation and impairment
50–100 ppmv	Characteristic eye damage called 'gas eye'
100–320 ppmv	Loss of sense of smell, nausea and increased lung irritation
320–530 ppmv	Lung damage and water accumulation
530–1000 ppmv	Shortage of breath, stimulation of respiratory centre, convulsions and chance of respiratory arrest
1000 ppmv and above	Immediate collapse and respiratory failure immediately before death

increased number of red blood cells and their volumes but the main clinical effect is irritation of the lung passages and the eyes which may leave the victim with pneumonia and conjunctivitis for some days afterwards (Table 6.2). Recovery from acute poisoning is usually without after-effect but some cases may show persistent clinical symptoms resulting from oxygen deprivation. Current Threshold Limit Values (or OELs) for maximum exposure of workers to hydrogen sulphide, set at 10 ppmv for 8-hour daily exposures, would appear to be very high because these levels are also known to cause eye irritation. Chronic effects are characterised by conjunctivitis, headaches, dizziness, diarrhoea and loss of weight. Early French medical writers speak of '*plomb des fosses*' to describe the colic and diarrhoea of Paris sewermen probably from such a cause which resembled that normally associated with lead poisoning.

6.5 HYDROGEN FLUORIDE AND FLUORIDE IONS

(a) Ubiquitous by-product

Fluorine is a gas so reactive that it does not occur naturally in elemental form but many common fluoride-containing minerals such as fluorspar, CaF_2, cyrolite, Na_3AlF_6, and certain apatites, e.g. $Ca_{10}F_2(PO_4)_6$, are used extensively by industry. Some industries produce gaseous hydrogen fluoride or its solution product hydrofluoric acid as by-products, but increasing sophistication of chemical industries has brought about an increase in the use of hydrogen fluoride to order to form various useful fluoro-derivatives. The variety of industries that now produce fluoride as a by-product or have a significant use of fluorine derivatives is shown in Table 6.3. These possible sources are superimposed upon significant natural background of sources of fluoride either dissolved in water or emitted from volcanoes. Thus levels in both air and water supplies may vary widely. The majority of rural and urban air monitoring sites record very low levels of atmospheric fluoride (i.e. total water-soluble fluoride) but, near phosphate fertiliser plants, aluminium smelters or volcanoes, levels may rise above 200 ppbv. Water supplies from these immediate areas may also show elevated levels, well above the 1 to 1.2 ppm recommended as an optimum set so as to provide an 'acceptable' incidence of dental caries and at the same time allow the correct bone growth of children. Food and drinks are the most important sources of human fluoride intake and normally these contain below 1 ppm of fluoride. Tea (3–180 ppm), rye (12–14 ppm), fish and other seafoods (2–85 ppm) are heavily laden exceptions. Other vegetables and

Table 6.3 Industrial and commercial processes involving fluorine compounds which may cause releases of fluoride and hydrogen fluoride to occur

Emission processes	Uses of fluorine derivatives
Aluminium smelting	Clouding of electrical bulbs
Steel production	Cut-glass finishing
Phosphate fertiliser production	Aviation fuel production
Enamel and pottery manufacture[†]	
Brick making	Separation of uranium isotopes
Missile propulsion[†]	Synthesis of plastics
Beryllium, zirconium, tantalum and niobium purification	Aerosol, refrigerant and lubricant manufacture
Cleaning of castings[†]	Wood preservation
Welding[†]	Cement reinforcing
Sandstone and marble cleaning[†]	Furniture cane bleaching
Cryolite, fluorspar and apatite mining	Supplementation of water supplies

[†] Enchanced by deliberate use of HF and other fluorine derivaties.

cereals grown in areas subjected to high fluoride emissions may also be enriched in fluoride by several orders of magnitude above normal.

Some legumes of the Southern Hemisphere (e.g. some *Acacia*) are unusual in that they accumulate fluoride as fluoroacetate which is over 500 times more toxic than inorganic fluoride and one of the most toxic simple molecules known. Fluoroacetate itself is not toxic until it is converted to fluorocitrate which inhibits the enzyme aconitase in the tricarboxylic acid cycle within mitochondria (see Appendix, Sect. 8.4). There is little evidence to suggest that cultivated legumes form fluoroacetate in the presence of excess fluoride although they do have the enzymes to cleave carbon-fluoride bonds.

(b) Accumulation by plants

The economic extent of crop damage due to fluoride is ranked fourth in importance after ozone, sulphur dioxide and nitrogen-based air pollutants in the USA. In weight-for-weight terms, fluoride is also the most phytotoxic of all atmospheric pollutants, with injuries to the most susceptible plants occurring at concentrations between 10 to 1000 times lower than caused by any of the other air pollutants. Actual rates of uptake of fluoride into leaves is also faster than any other pollutant and these accumulations may also cause problems to animals feeding upon such fluoride-rich crops (see later).

Both gaseous and particulate fluorides are deposited on plant surfaces, and some of this may penetrate if the leaf is old or weathered. Nevertheless, the main access into a plant is by gaseous fluoride entering through the stomata. An important feature of fluoride uptake and transport in plants is that it is carried in the transpiration stream towards the leaf tips or margins where it accumulates to high concentrations and this is where phytotoxic effects usually develop. Normally, soils contain between 20 to 500 μg of fluoride per gram but, because fluoride has limited solubility in soilwater, uptake by roots is relatively low and there is little relationship between soil fluoride and total plant fluoride content. Thus atmospheric sources of fluoride are more important than fluoride in groundwater in determining the amount of fluoride in or on a crop. This problem is well known in relationship to the occurrence of fluorosis in farm animals. Animals grazing on pasture very close to brickworks, smelters and phosphate fertiliser factories, or fed forage gathered from such areas, may be subjected to excessive intakes of fluoride which may lead to fluorosis, a condition also occasionally found in humans (see later). The major recommendation (known as the 'Washington' Standard) to avoid this problem has been to ensure that the yearly average fluoride content of herbage does not exceed 40 μg g^{-1}.

There are several mechanisms which lead to a reduction in fluoride content of plants which include shedding of individual leaves or surface waxes, leaching by rain or volatilisation. Fluoride levels are often lowest during summer months for a variety of reasons; partly because of more favourable meterological conditions for better dispersal of local sources of fluoride pollution and partly due to greater turnover of leaves in grass swards during summer.

Different species show wide ranges of susceptibilities to fluoride. Young conifers, gladioli, peaches (which show a defect called 'suture red spot') and vines are especially sensitive whilst tea or cotton are unusually resistant. Environmental factors, such as light, temperature, humidity, water stress, etc., all influence plant response. There are many descriptions of different types of metabolism such as photosynthesis, respiration or metabolism of amino-acids, proteins, fatty acids, lipids, and carbohydrates being adversely affected by presence of fluoride (see Table 6.4 for details) but adequate explanations of exactly why these are inhibited are still not available. Certain enzymes are modulated by the presence or absence of fluoride (e.g. the enolase of the tricarboxylic acid cycle) but these cannot explain the wide range of metabolic changes known to occur. The clue most likely to provide an overall explanation resides in the interactions that take place between fluoride and calcium or magnesium. Calcium and fluoride, for example, together stimulate phosphate uptake by potato tuber discs, a process that is temperature independent. This means that the fluoride effect is non-metabolic and that, possibly, calcium

Table 6.4 Physiological effects of fluoride on plants

Process	Disturbance	Likely cation interaction
Respiration and carbohydrate metabolism	Glycolysis inhibited	Mg^{2+}
	Pentose phosphate pathway enchanced	Mg^{2+}
	Unusual mitochondrial swelling	Mg^{2+}
	Oxidative phosphorylation reduced	Mg^{2+}
Photosynthesis	Unusual chloroplast structure	Mg^{2+}
	Inhibited pigment synthesis	Mg^{2+}
	Increased PEPC[†] activity	Mg^{2+}
	Reduced electron flow	Ca^{2+}
Amino acid and protein metabolism	Increase in free amino acids and asparagine	Ca^{2+}/Mg^{2+}
	Decrease in ribosome sizes	Ca^{2+}/Mg^{2+}
Nucleic acid metabolism	Changes in transcription and translation	Ca^{2+}/Mg^{2+}
Fatty acid and lipid metabolism	Increased esterase activities	Ca^{2+}/Mg^{2+}
	Decreased unsaturated/saturated ratios	Ca^{2+}/Mg^{2+}
Other metabolic changes	Increase in peroxidase activity	Ca^{2+}/Mg^{2+}
	Decrease in acid phosphatase activity	Ca^{2+}/Mg^{2+}
Transport and translocation	Altered plasma membrane ATPases	Ca^{2+}
Fruit development	Poor fertilisation and seed germination	Ca^{2+}
	Reduced pollen tube growth	Ca^{2+}
	Reduced seed number and fruit size	Ca^{2+}

† Phospho-enolpyruvate carboxylase activity as in C_4 photosynthesis (see Sect. 6.2c).

adsorption sites on the cell walls and the plasma membranes may be implicated in the response to fluoride. Cytoplasmic calcium is an ubiquitous regulator of cell metabolism and many, but not all, of its effects are mediated by a calcium-binding protein calmodulin, which in turn stimulates a variety of enzymes including calcium-dependent protein kinases. Furthermore, calcium ions are known to affect the transport selectivity of membranes with respect to other substances. Because of this interaction with calcium, fluoride will naturally exert a more general effect on various cellular regulatory activities (as listed in Table 6.4) rather than have recognised individual sites of damage, as with so many other atmospheric pollutants. This probably also explains why it is so phytotoxic at such low concentrations.

Fluoride also forms magnesium-fluorophosphate complexes and consequently many enzyme pathways (see Table 6.4) are adversely affected by fluoride. Most reactions involving ATP require additional magnesium complexes in order to function correctly. If these natural complexes are disturbed by the presence of additional fluoride then other key reactions may also be inhibited.

Application of lime to crops and herbage has long been known to be a practical means of reducing the effects of fluoride injury. Originally, it was thought the lime caused the immobilisation of the fluoride on the surfaces of the leaves as insoluble calcium fluoride, CaF_2. However, calcium chloride spraying has a similar alleviating effect to lime and recent studies have shown that the remedy actually relies upon additional calcium entering the leaves to interact with the fluoride inside and redress any calcium imbalances in the regulatory processes. Had more thought been given to this simple procedure to alleviate fluoride injury within plants, the relationship of fluoride to calcium (and magnesium) levels within the tissues would have been appreciated much earlier and progress that much greater by now.

(c) Fluorosis in animals

The US Department of Agriculture claims that fluorides 'have caused more world-wide damage to domestic animals than any other air pollutant'. The Icelanders have good cause to agree with this. The eruption of the Icelandic volcano Hekla in 1970 caused problems in animals within a 130-mile radius and killed over 7500 sheep and lambs by fluorosis (chronic fluoride poisoning). Historical records showed that this was not an isolated event, similar problems with animals occurred after the volcano had erupted 200 and 900 years before. Such acute episodes are far less common than chronic intoxication in sheep and cattle caused by eating fluoride-rich grass, hay or silage over long periods. Inevitably, contaminated soil is also taken in as the animals graze. The symptoms are manifested in two stages in cattle, a lethargic phase when hides become less elastic and the animals show clear signs of pain in the rib region when pressed, followed by a lameness phase with swellings and loss of milk production which leads to a rigidity of the back bones, pain in the legs and, finally, death by wasting. Pigs and sheep show a similar sequence of disturbances and it is thought that wild animals may also show fluorosis. The problem is not just confined to mammals; it is impossible to raise silkworms in the vicinity of fluoride emissions whilst fluoride has replaced arsenic as the most serious cause of death to bees. Levels of fluoride in insects from fluoride-affected areas may rise as high as 400 ppmv, well above average levels of 8 ppmv, with proportional differences in death rates.

(d) Human fluorosis, fluoridation and dental health

As already mentioned, food and drink are the main avenue of fluoride uptake for the population at large, but workers in fluoride-generating industries (particularly cryolite mining) breathe in fumes and dusts containing fluoride which are then absorbed into the body through the respiratory tracts rather than being taken in through the digestion systems. The Threshold Limit Value (or OEL) for an 8-hour working day is set at 3 ppm for hydrogen fluoride but, because fluoride is lost in the urine, elimination rates of below 4 mg of fluoride per day are used as a monitor of safe working conditions. Storage of fluorides takes place principally in the bones and teeth with normal ranges being from 50–500 ppm, but these levels may rise dramatically within tea drinkers living in areas supplied with naturally fluoride-rich waters.

Understanding at the biochemical level of the toxicity of fluoride in humans is fragmentary, as it is in plants. The high affinity of fluoride for magnesium, manganese, iron, calcium and phosphate causes it to interfere with many enzymes (see Table 6.4) and to adversely affect the function of hormonally controlled processes. The parathyroid function which regulates the metabolism of calcium appears to be counteracted by raised fluoride intakes. This, as well as other effects, is largely due to the sequestering effect of newly forming bone tissues which take up fluoride and thereby influence the balance between carbonate and calcium. The variety of processes in humans affected by fluoride (Table 6.5) mirrors the range of disturbances found in plant tissues (Table 6.4). By extrapolation, it is more than likely that such changes in animals are brought about by the influence of fluoride upon homeostasis associated with membrane-associated events involving calmodulin and calcium-dependent protein kinases or unusual magnesium fluorophosphate complexes being formed to prevent normal magnesium and phosphate activation of enzymes.

The symptoms of fluorosis in man are broadly similar to those in animals although acute cases are very rare. Dental fluorosis is characterised by permanent, white or chalky patches on the dental enamel where there has been imperfect calcification and a deficiency of cementing substance. In the advanced stage of chronic fluorosis in

Table 6.5 Physiological effects of fluoride in animals and humans

Process	Disturbance
Carbohydrate metabolism	Glycogen levels depleted Glycogen turnover depressed Phosphorylase reduced
Lipid metabolism	Activation of acetate inhibited Liver lipases activated Certain esterases inhibited
Mineral metabolism	Interference in iron uptake Sulphite and phosphate counteract the inhibiting effect of calcium upon intestinal absorption
Hormonal balances	Effect on parathyroid function[†]

[†] Calcium levels are influenced by parathroid hormone (parathormone, PTH) produced by the parathyroid and a hormonal derivative of vitamin D (called 1,25-dihydroxycholecalciferol) found in the liver and kidneys (both of which compounds raise blood serum levels of calcium) and calcitonin released by the thyroid which causes enhanced calcification of the bone tissues and therefore reduces blood calcium levels.

humans, which are very infrequent, the mottling may become yellow, red, brown or black whilst the tissues and ligaments may also become calcified and protrusions may develop on the surfaces of bones. Sometimes these can be so pronounced that they bring pressure upon spinal nerves and cause paralysis. Rigidity of the rib-cage may also cause breathing difficulties and a variety of other symptoms may arise such as arthritis, muscle pains and migraine. The occurrence over the body, especially the limbs, of small pinkish-brown areas which resemble bruises and called 'Chizzola maculae' has been recognised as being associated with fluorosis. They clear up when sources of fluoride contamination (air, water or food) are removed or when sufferers take a rest from work, or when they leave the vicinity of fluoride emissions and re-occur when fluoride is encountered again.

It is largely problems associated with too much fluoride that have coloured arguments over fluoridation of public water supplies. Signs of fluorosis occur when fluoride exceeds 1.4–1.6 ppm in drinking water. On the other hand, the incidence of dental caries in children from areas supplied with water with fluoride lower than 0.5 ppm is between two and four times higher than those having 1 ppm fluoride in their drinking water (see Fig. 6.8). Medical and public authorities have sought to optimise the lesser of one evil against another by advocating fluoridation of water supplies to a level of 1 ppm.

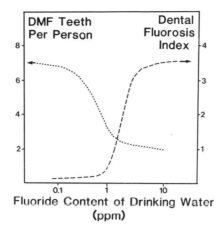

Fig. 6.8 The balance between the incidence of dental decay (where DMF is the number of decayed or missing or filled teeth) and signs of dental fluorosis as a consequence of fluoride in the drinking water.

SELECTED BIBLIOGRAPHY

Binder, K. and Hohenegger, M. (eds.), *Fluoride Metabolism.* Verlag Wilhelm Maudrich, 1982, Vienna, Munich and Berlin.

Coburn, R. F. (ed.), Biological effects of carbon monoxide. *Annals of the New York Academy of Sciences* **174**, 1970, article I, 1–430.

Filler, R. and Kobayashi, Y. (eds.), *Biomedicinal Aspects of Fluorine Chemistry.* Kodansha 1982, Tokyo, and Elsevier Biomedical Press, Amsterdam.

Martensen, L. M. (ed.), Symposium on CO_2 enrichment. *Acta Horticulturae* **162**, 1984, Aas, Norway.

Murray, F. (ed.) *Fluoride Emissions – their Monitoring and Effects on Vegetation and Ecosystems.* Academic Press, 1982, Sydney, New York and London.

Royal College of Physicians, *Fluoride, Teeth and Health.* Pitman Medical, 1976, Tunbridge Wells.

Weinstein, L. M., Fluoride and plant life. *Journal of Occupational Medicine* **19**, 1977, 49–78.

World Health Organisation, *Fluorides and Human Health.* 1970, Geneva.

7

Interactions, 'recent forest decline' and other considerations

'Science moves, but slowly slowly,
creeping on from point to point.'

From 'Locksley Hall' by Alfred, Lord Tennyson (1809–1892) published in 1842.

'In research, the horizon recedes as we advance . . .
and is no nearer at sixty than it was at twenty.
The urgency of the pursuit grows more intense . . .
and research is always incomplete.'

From the biography of 'Isaac Casaubon' by Mark Pattison (1813–1884) published in 1875.

7.1. INTERACTIONS

(a) Uncertain contributions of stress

The thoughts encapsulated by the two contemporaries above express the sentiments of many who investigate some of the problems that have arisen as a consequence of environmental change brought about, at least in part, by air pollution. Earlier chapters have already identified individual pollutants that are known to be major causes of specific problems in plants or animals or with materials. Even in the case of 'acid rain' (a diverse combination of the acidic products of air pol-

Table 7.1 Different types of plant stress

Stress	Type	Variations
Physicochemical (abiotic)	Temperature	Low (frost and winter cold or chilling) High (heat)
	Water (including humidity)	Deficit (drought) Excess (flooding)
	Radiation (intensity and quality)	Infrared, visible and ultraviolet Ionising
	Chemical	Salt (sea-spray, road salting and hydrological change) Atmospheric gases, acid rain and particulates Herbicides, insecticides, etc.
	Others	Seasonal, wind pressure, sound, magnetic, electrical
Biotic	Animalia[†] (mainly insects, grazing herbivores or nematodes) Fungi[†] (mildew, etc) Higher plants or Plantae[†] (parasitic plants like mistletoe) Protista[†] (unicellular eukaryotes including protozoans, yeasts, etc.) Monera[†] or procaryotes (bacteria, etc.) Viral	
Edaphic[‡]	Vital trace element availability Heavy metal toxicity Mycorrhizal associations Compaction, weathering	

[†] Assumes the modern five-kingdom classification system of Whittaker.
[‡] Stresses associated with soils which may also include some of the abiotic and biotic stresses above.

lution in solution), adverse changes in water quality brought about by acidic precipitation are now known to be the major cause of the decline in freshwater fish stocks in certain sensitive areas. Nevertheless, there are a number of environmental problems, largely associated with detrimental changes of vegetation, that cannot be attributed directly to a single gaseous pollutant or to acidic precipitation, even just to air pollution.

Any environmental factor that is likely to be unfavourable to living organisms is popularly called 'stress' (although it really should be called 'distress' if one is using the word properly) and the stress resistance of an organism is defined as the ability to survive unfavourable stress. Such stresses are often subdivided into biotic and abiotic types to distinguish those caused by other biological organisms from those of a non-living physicochemical nature. In Table 7.1 an attempt is made to categorise the most important plant stresses.

(b)   Synergistic or more-than-additive?

Clearly, with so many different types of stress, it is often the case that some of them occur simultaneously or immediately after each other. As a result, the outcome is less predictable than with a single stress. For example, two stresses may have distinctly different modes of action which have no effect upon each other and they simply go on causing the appropriate amount of change which when added together are equal to the sum of the two separately. Such an effect is known as an 'additive' change and is not an interaction. If, however, the sum of the two effects is less than those caused by the two individual stresses separately, the interaction is known as 'antagonistic'. In other words, one stress apparently alleviates the effect of the other but if the sum of the combined effects is greater than the two separate effects added together then the interaction is said by some to be 'synergistic'. The exact definition and the proof that 'synergism' has taken place often gives rise to controversy because the word can so easily be misused. The use of the term 'more-than-additive' instead is probably better in most instances.

The real problem lies in defining and determining how relative amounts of stress are measured. In the case of air pollution, each pollutant can be measured in terms of concentration (ppbv or μg m^{-3}) or of dosage, (ppbv h^{-1} or μg m^{-3} h^{-1}) and is best illustrated by consideration of a hypothetical experiment. If, for example, a group of plants, S, was notionally exposed for 72 hours to a concentration of 50 ppbv of sulphur dioxide, another similar group, N, to 50 ppbv of nitrogen dioxide, yet another, S + N, to a mixture of 50 ppbv of sulphur dioxide and 50 ppbv of nitrogen dioxide simultaneously and then at the end of the fumigation the growth of all three groups, S, N, S + N, compared to growth in 'clean' air control plants. Should the depression of growth in group S turn out to be 20 %, group N, 10 %, and that of group S + N, 60 %, then suitable statistical tests, would probably show (if enough replicates were undertaken) that an interaction has taken place which some would claim is 'synergistic'. The main difficulty with this experiment, leaving aside the problem of what is a 'clean' air control (see Sect. 2.2e), is deciding if the treatment S + N using 50 ppbv sulphur dioxide plus 50 ppbv nitrogen dioxide for 72 hours is the correct mixed treatment for this experiment. The concentration (or dosage) has actually been doubled and by doing this one may have passed a threshold which changes plant response in a different manner. Perhaps one should have used 25 ppbv sulphur dioxide and 25 ppbv nitrogen dioxide, S/2 + N/2, that is an equivalent concentration or dosage to S and N combined as the mixed treatment instead? My own view (probably heretical) is that, in any thorough experimentation to determine interactions, two mixed treatments are required. In this case, S + N and S/2 + N/2, plus S and N along with two controls, 'clean-air' (C_A) and 'charcoal-filtered' (C_F) as discussed in Sect. 2.2(e). If all four tests, S + N, S and C_A; S + N, N, S and C_F; S/2 + N/2, N, S and C_A; S/2 + N/2, N, S and C_F, show significant interactions then 'synergism' between two pollutants exists. However, if only some of the statistical tests are significant then only the use of the term 'more-than-additive' is justified. This may be an acceptable compromise between the two viewpoints. Thus synergism takes final concentration or dosage of pollutant into account as an entity in its own right whilst in the latter (i.e. more-than-additive) the treatment with different pollutant gases is all-important.

This pedantic approach may be only applicable to comparisons between 'similar' types of stress (in this case two or more air pollutants) because both the chemical natures and the ultimate concentrations of air pollutants are involved. In the study of an interaction between, for example, sulphur dioxide and drought stress, the concentration (or dosage) of the pollutant remains the same in the single and mixed treatments. Consequently, a distinction between more-than-additive or synergistic interaction (if it is demonstrated by statistical analysis) is not necessary. It also demonstrates the problems of using concepts developed in one area of study and transferring them to another without critically thinking them through their new relationships each time.

The above example also illustrates another point made much earlier in Ch. 1 (Sect. 1.2) about the choice of units to express concentrations of atmospheric pollutants. By using ppbv expressions for both sulphur dioxide and nitrogen dioxide, their direct comparability on a molecule-to-molecule basis at any temperature can be appreciated. If, however, each concentration is expressed as mass per unit volume (i.e. 131 μg m^{-3} sulphur dioxide and 94 μg m^{-3} nitrogen dioxide only at 25°C and 101.3 kPa) the equivalence is not immediately clear. Such ease of comparison in mixed treatments is the *only* justification for the retention of non-standard units of measurement such as ppbv or ppmv.

(c) Effects of mixed pollutants on plants

Until the start of this decade, research interest in Europe concerning the effects of ambient air pollutants on plants concentrated heavily upon sulphur dioxide whilst in North America rather more interest was paid to the effects of ozone. These different approaches were appropriate because, prior to the 1980s, such pollutants were recognised to be associated with the major pollution problems in these particular areas. Since then Europeans have realised that oxides of nitrogen may accompany sulphur dioxide at similar or even greater concentrations and, under suitable meteorological conditions, conditions for ozone formation may be created almost anywhere in Europe. Ironically, the success with which legislation has reduced the levels of particulates and smoke from the atmosphere over Europe has now extended those areas where ozone is more likely to be formed. In North America, the co-occurrence of sulphur dioxide and oxides of nitrogen emissions is now more widely recognised although it is still believed that ozone is generated some distance from the sources of oxides of nitrogen and a combination of all three is much less likely than in Europe.

In Ch. 5, the daily pattern of photochemical smog formation is described in some detail and a typical example, characteristic of this daily phenomenon, is illustrated in Fig. 5.3. The main feature of photochemical smog formation is that little or no sulphur dioxide is emitted and it requires several non-windy, warm, bright days in succession to become established. In many industrial parts of the world these conditions are infrequent and cyclical daily patterns such as that shown in Fig. 7.1 are rather more common especially in Europe. In such cases, levels of photochemical oxidants are usually lower (and occur mainly later in the day) but diurnal emissions of sulphur dioxide for heating and power generation coincide with the transport of people to work in the morning (when oxides of nitrogen are also emitted). Emissions of sulphur dioxide continue as industrial and commercial activities take place, decline as the workers return home (with a second rise in the levels of oxides of nitrogen) and then later rise as they need more heat and power before going to bed. By contrast, ozone formation is highest in the middle to late afternoon.

Uptake of pollutants into plants is mainly through their stomata (see Ch. 2, 3 and 5); notwithstanding the possible exception of a proportion of the oxides of nitrogen (see Ch. 3) entering through the cuticles. This means that in urban and semi-urban environments at

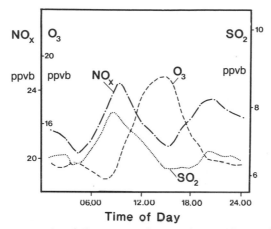

Fig. 7.1 Representative daily pattern of atmospheric pollutant fluctuations more typical of general pattern than Fig. 5.3. (Courtesy of Drs Nicholson, Fowler, Paterson, Cape and Kinnaird of the Institute of Terrestrial Ecology, Scotland.)

least, those atmospheres most likely to affect plants are those prevailing between 8 a.m. and 4 p.m. which have their greatest influence at midday when light intensities and gaseous fluxes are highest. In the case of deciduous plants, the time when leaves are present and the periods of most active growth would normally be regarded as the most sensitive seasons, although in the case of evergreens (conifers, grasses, etc.) this period of sensitivity may be longer. Nevertheless, these are only broad generalisations and exceptions to the rule are common.

Furthermore, it has always been difficult to reconcile the experimental observations of smaller reductions of growth of a particular plant, caused by experimental fumigation using a single pollutant at a level normally found in semi-urban environments as compared to clean air, to the larger reductions in the growth of similar plants grown in ambient (semi-urban) air when compared with those grown in charcoal-filtered air. The answer is partly one of 'which is the real control?' discussed in Ch. 2 and partly the possibility of interactions between air pollutants taking place.

Most research studies of interactions concerning air pollution involve just two pollutants; usually sulphur dioxide and ozone or sulphur dioxide and nitrogen dioxide. Studies of combinations of nitrogen dioxide with ozone or, intentionally, mixtures of nitrogen dioxide with nitric oxide, are less frequent and the numbers of studies involving three or more pollutants simultaneously are very small indeed. This is not altogether a surprise when double-pollutant studies are more than twice as difficult to undertake as single-pollutant studies and three-way pollutant studies have at least fourfold the technical complexity.

The results from investigations of possible interactions now reported in the literature reveal a wide and bewildering array of additive, more-than-additive, or antagonistic effects. This is to be expected because each study has a choice of different pollutant, concentrations, duration of exposure and fumigation conditions using various species, cultivars or clones, and often measure quite different parameters of change (i.e. dry or fresh weight gain or loss, grain yield, leaf injury, wet photosynthesis, enzymatic change, etc.). Rather than bury the salient observations from these various studies in a mass of detail, Table 7.2 summarises a number of the most important inter-actions that frequently occur in the presence of mixed pollutants. Once again these should be treated only as a guide. Exceptions always exist and the various permutations of treatment, species and parameters are always the cause of uncertainty. In my opinion, such studies of interactions will always add to knowledge and some of this is very useful but, unless these investigations are accompanied by attempts to understand the underlying mechanisms, no real progress will be made. An alternative approach is to attempt to understand more fully the variations in the chemistry, biochemistry and physiology associated with different single pollutants as has been outlined in previous chapters. Another is to develop a model which if it is sufficiently robust will be adaptable and predictive. Such ideas are developed further in the last section of this chapter. The third, probably a combination of the other two, is best called the 'inspired guess' approach in which the few clues showing that natural metabolism or homeostasis in the presence of one pollutant may differ in the presence of another are examined more closely. This type of approach has had some success (see below).

(d) Mechanisms of pollutant interaction

One unusual characteristic of sulphur dioxide atmospheric pollution is that at certain humidities it causes the stomata of some species to open wider than would normally occur in 'clean' air. This effect persists down to concentrations as low as 19 ppbv (and high as 570 ppbv) and means that more sulphur dioxide (and carbon dioxide) gains access to leaf cells (and more water is lost) than normally would be the case. This opening effect is greatest at high humidities. If the humidity is low then the effect of sulphur dioxide is one of closure instead of opening and it is thought that nitrogen dioxide causes similar opening at high humidities, but little research has actually been done on the detailed effects of this gas alone on stomata. By contrast,

Table 7.2 Known interactions caused by atmospheric pollutant mixtures

Mixture	Interaction	Characteristic affected
$SO_2 + O_3$	Additive[†]	Leaf growth (radish)
	More-than-additive	Leaf appearance (grapes, radish, peas, tobacco, pines)
		Root growth (tobacco)
		Plant weight (soyabean)
	Less-than-additive	Fungal infection
		Leaf appearance (apple, soyabean, tomato)
		Plant weight (alfalfa)
		Needle length (pines)
$SO_2 + NO_2$	Additive[†]	Foliar injury (many species)
	More-than-additive	Pollen tube growth
		Overall yield (especially grasses)
		Enzyme activity (peas)
$NO_2 + O_3$	More-than-additive	Pollen tube growth
	Less-than-additive	Overall yield
$SO_2 + NO_2 + O_3$	More-than-additive	Premature leaf fall (poplar)
		Leaf appearance (radish and peas)
$SO_2 + HF$	More-than-additive	Leaf appearance (only barley and maize)
$O_3 + HF$	More-than-additive	Leaf appearance (only *Coleus* and mint)
$NO_2 + NO$	Additive[†]	Photosynthesis (tomato)
$O_3 + H_2S$	Less-than-additive	Photosynthesis (barley, maize and citrus)

[†] An additive effect is not strictly an interaction when strictly defined.

mixtures of sulphur dioxide and nitrogen dioxide (115 ppbv each) promote closure of stomata in conditions under which sulphur dioxide alone would cause opening. Such an important physiological difference involving the mechanisms by which vegetation actually interfaces with the atmosphere is bound to cause a major interactive effect upon the parameters governing net growth and photosynthesis, especially those components that depend on precise control of water content.

By contrast, ozone does not have a specific and direct effect on stomatal opening but may interact with other gases or stresses and accentuate their effects because the effect of this gas is primarily upon membranes and their degree of 'leakiness' to water and important ions. As has already been mentioned in Ch. 5, ozone effects are accentuated in those membranes, particularly plasma membranes, in those cells which have high transport responsibilities such as those loading and unloading transport vessels (i.e. the Strasburger or albuminous cells of conifers) as well as the guard cells of stomata. Because of what is known about stomatal function and the fact that guard cells are especially sensitive to combinations of sulphur dioxide, nitrogen dioxide and humidity (as well as a number of other environmental and internal factors involved in control of water content), it is very likely that osmotic and ionic disturbances are a part of such interactions as well as those caused unspecifically by ozone and that any mechanism to explain 'interactions' should take them into account.

Another similar clue is provided by studies using the transmission electron microscope to determine the ultrastructural changes at high magnification caused by sulphur dioxide, oxides of nitrogen and ozone. In each case, one of the first symptoms of each of the individual gases is the observation of swelling of the spaces inside the thylakoids (see Appendix, Sect. 8.3) within the chloroplasts where many of the reactions of photosynthesis take place. It is known that pH gradients naturally exist across the thylakoids during bioenergetic conversion of energy and during specimen processing for transmission electron microscopy severe osmotic changes are always involved. It would appear that the presence of the pollutants disturbs the pH balance, or sensitises the tissues, so that subsequent osmotic disturbances are greater or that a combination of the two disturbances coexist.

If, under certain conditions both sulphur dioxide and nitrogen dioxide close stomata, they may directly inhibit overall photosynthesis but it is also possible that the products of sulphur dioxide, nitrogen dioxide or ozone in solution (e.g. sulphite, hydroxyl radicals, etc.) might adversely affect components of the photosynthetic mechanisms in both the thylakoids or in the surrounding components of the stroma. Detailed *in vitro* studies on the effects of sulphite or nitrite upon such components have so far failed to reveal any evidence of sensitivity which could explain these depressions of photosynthesis when mixtures of gases each at similarly low levels together cause quite obvious reductions in net growth. However, studies of the pH gradients built up across the photosynthetic membranes (thylakoids) from the alkaline stroma to the acidic lumen inside the thylakoids (see Figs. 8.2 and 8.3) have shown a breakdown of pH gradients in the combined presence of small quantities (0.1 mM) of sulphite and nitrite ions (some of the products of sulphur dioxide and nitrogen dioxide in solution) or after small pulses of ozone. By contrast, if sulphite or nitrite were provided separately even at much higher concentrations (1 mM) no similar deterioration of pH gradients occurred. Moreover, the characteristics of the destruction (and later repair) of the pH gradients in the presence of mixtures of sulphite and nitrite or ozone were exactly those that would be caused by free-radical action (see Appendix, Sect. 8.2). This probably means that small amounts of sulphite and nitrite together inside biological tissues are capable of producing free radicals (e.g. $HSO_3^{\bullet}$) where separately they would be unable to do so, and these active species act upon membranes causing them to lose some of their vital properties. Expressed another way, they are capable of rendering membranes more susceptible to a 'leakage' of protons back across the membrane where their normal function would be to hold apart differences in levels of acidity within the cell. This separation is essential if vital bioenergetic functions (e.g. ATP formation in chloroplasts and mitochondria, see Appendix, Sects. 8.3 and 8.4) are to take place efficiently. If such 'leaks' are caused by mixtures of sulphite and nitrite or ozone this means that less ATP might be formed or more energy (in the form of light or sugar) needs to be put in to create a similar amount of ATP and, as a consequence, less energy is available for growth. Furthermore, cellular energy is also required to repair damage done by free-radical action and again this is unavailable for growth, reproduction, etc., if one accepts the 'energy-diversion hypothesis' introduced earlier (Sect. 1.5).

This osmotic–ionic mechanism involving free radicals is, so far, the best explanation of the more-than-additive effects of sulphur dioxide, nitrogen dioxide and ozone upon plant growth. Other mechanisms explain the experimental facts less well but none of them have the full range of characteristics that fit the few clues available.

One interesting observation has been that plants normally increase the catalytic capacity of the enzyme nitrite reductase in order to convert nitrite into ammonia (see Sect. 3.2c) in response to nitrogen dioxide in the atmosphere, whereas sulphur dioxide alone has no effect upon this enzyme. This induction of nitrite reductase is one of the means by which the plant detoxifies, and even takes advantage of, the nitrite caused by nitrogen dioxide pollution. However, when both sulphur dioxide and nitrogen dioxide are present simultaneously, the induction of nitrite reductase does not occur. This means that such plants experiencing both pollutants are partially prevented from adjusting their metabolism so that extra nitrite cannot be easily detoxified. In fact, they are doubly hampered because not only do they have to cope with the toxicity associated with sulphite but also with additional nitrite which is also highly toxic to plants. At the moment, it is believed that this explanation of an interaction between sulphur dioxide and nitrogen dioxide is less important than one based on enhanced levels of free-radical activity within membranes. Should mechanisms of both types be subsequently found to be superimposed upon each other then it will be more difficult to find their exact relationships but we will be closer to an explanation of why and how interactions do occur.

(e)   Do 'cocktails' affect humans?

Earlier parts of this book have shown that, with the certain exception of carbon monoxide and the possible exception of hydrogen sulphide, plants are generally more sensitive to atmospheric pollution than animals. When the basic modes of action are examined in each case, the similarities in molecular detail are greater than the differences. Very often, the latter distinctions are associated with specific functions such as photosynthesis in plants or local hormone and immunological responses in animals.

Previous sections of this chapter have also summarised the wealth of detail that exists, largely from the study of plants, to show that interactions between different atmospheric pollutants and/or different stresses are extremely important aspects of the study of air pollution and cannot be ignored. In studies of animals and health, however, there is no similar background of information concerning the possibility of interactions taking place despite the fact that plants and animals (and people) inhabit the same polluted environment. Rational explanations are easily provided. Studies of human health in the presence of atmospheric pollution are usually carried out by extensive epidemiological surveys, including those of medical records within critical localities. For more basic mechanistic studies, uses of experimental animals (with their attendant problems of extrapolation to the human condition) are difficult enough with one pollutant and do not usually permit sensible studies of mixtures of different pollutants.

Other chapters have already made the point that in making accurate epidemiological surveys, it is difficult to account adequately for the effects of smoking, poverty and other variables away from those of atmospheric pollution. Nevertheless, there is now a growing awareness amongst medical statisticians that, in some areas, there are higher rates of infant mortality and more cases of bronchitis than one pollutant acting alone could be expected to cause. Most of them now recognise that mixtures (or a 'cocktail' as they call it) of pollutants may be causing problems with human health on a larger scale than previously thought possible. In a sense the EEC Limit Values (Table 1.3) are the first recognition of this possibility in tangible form. Atmospheric levels of sulphur dioxide are now modulated within the EEC by the level of smoke that exists simultaneously. If the experience of plant scientists over the past 15 years with interactions is heeded and subsequently taken into account, then international guidelines of air quality might well develop so that harmful coincidences of potentially harmful concentrations of sulphur dioxide, smoke, acidic precipitation, oxides of nitrogen and ozone are avoided.

7.2 'RECENT FOREST DECLINE' – NEUARTIGE WALDSCHÄDEN

(a) Occurrence and classification

Over recent years, symptoms of a new type of damage to forest trees has occurred which has been given several different names. The most cautious is 'forest decline' used in the UK whereas 'forest' or 'tree die-back syndrome' is more frequently used in the USA. The Germans originally called it *'neuartige Waldschäden'* meaning new or novel kind of forest damage, which is probably the best description, but to emphasise the seriousness of the problem some writers have converted this phrase to *'Waldsterben'* or 'forest death'. I shall use the term 'recent forest decline' as meaning *neuartige Waldschäden*.

This unexplained phenomenon first appeared in West Germany in the early 1970s; initially on silver fir, *Abies alba*, and then on Norway spruce, *Picea abies*. Symptoms in the two are similar but it has been claimed that effects on Scots pine, beech and oak are caused by similar agents. The extent of the problem is illustrated by the survey data relating to Western European countries (Table 7.3) although it is widely believed that the adjacent Eastern European countries have similar (or worse) problems with the health of their forests. Difficulties may be also developing in the north-east states of the USA and the eastern provinces of Canada with red spruce, *Picea rubens*, and other major forest species.

Until the existence of the problem was widely accepted, surveys from forest to forest, area to area and country to country could not be correlated because different methods and assessments of the degree of damage were in use. However, a standard classification of the damage on a scale ranging from 0 (healthy) to 4 (dead) has become popular and has been advocated for use by member countries within the European Community. The essential details of this classification system are listed in Table 7.4 although considerable variations exist between tree species because their shape, growth characteristics and symptoms differ.

If one species of tree, e.g. Norway spruce, is taken as an illustrative example of the problems of classification, there is a great danger of being highly selective over the choice of pictures and thereby leading to the criticism of over-dramatising the problem. The selection of

Table 7.3 Reported percentage of different tree species affected by 'recent forest decline' in Western European countries

Tree species	Country (year)							
	W. Germany			E. France	Switzerland	Austria	Italy (S. Tyrol)
	(1982)	(1983)	(1984)	(1985)	(1984)	(1984)	(1984)	(1984)
Norway spruce	9	41	51	52	16	11	29	16
Silver fir	60	75	87	87	26	13	28	35
Scots pine	5	44	59	58	17	18	30	6
Beech	4	26	50	55	3	8	—	—
Oak	4	15	43	55	4	9	—	—
Others	4	17	31	31	6	9	—	—

Table 7.4 System of classification of trees according to health as recommended by the European Community

Class	Needle (or leaf) loss (%)	Description
0	0–10†	Healthy
1	11–25‡	Slightly damaged
2	26–60‡	Medium to seriously damaged
3	61–99	Dying
4	100	Dead

† Where existing needles (or leaves) between 26–60 % are yellowed then one class higher, above this two classes higher.

‡ Where over 25 % of existing needles (or leaves) are yellowed, one class higher.

Note: Optional parameters which may be included during classification include crown deformation, trunk condition, annual terminal shoot length, age of oldest needles, fall of healthy twigs, extra twigs, etc.

examples shown in Fig. 7.2 is probably as fair as any and comes from an information booklet for the general public entitled *Wald in Gefahr* (*Forest in Danger*) issued by the Bavarian State Government. It illustrates healthy (Class 0, top left), injured (Class 1, top right), damaged (Class 2, lower left) and heavily damaged (Class 3) Norway spruce trees showing progressive needle loss. It also illustrates (see top right) a feature of certain forms of this tree, the so-called comb-type, which when suffering from 'recent forest decline' exhibits a distinctive pendulous or streamer-like hanging of the secondary branches which becomes less obvious once a large number of needles are lost.

'Recent forest decline' symptoms, such as needle loss in Norway spruce, are often accompanied by yellowing of the needles which is allowed for in the classification system (see Table 7.4) but one characteristic of this yellowing is that the needles on the upper surfaces of the branches are more affected than those underneath, as though the 'problem' were coming from above. There are sometimes fresh replacement shoots on the upper surfaces as a secondary response to needle loss and yellowing.

The main problem with 'recent forest decline' and the classification of the degree of damage is that many other causes of damage coexist in the forest and it is far from easy to distinguish an 'unknown' cause of tree injury from that caused by a known stress or pathogenic agent. Indeed this difficulty has led to controversy even amongst experienced foresters, some of whom ask the question: 'Is recent forest decline a

Fig. 7.2 Series of Norway spruce trees showing the typical features associated with 'recent forest decline'. Top left, undamaged (Class 0); top right slightly damaged (Class 1) showing typical droopy 'comb' symptom; bottom left, moderately damaged (Class 2); bottom right, severely damaged (Classes 3 and 4). (Plates courtesy of the Bayerische Staatsministerium für Erhnährung, Landwirtschaft und Forsten, Munich, West Germany.)

real and fresh problem?'. On balance, when the surveys have been done carefully and all the effects of other agents have been eliminated, the answer is: 'Yes, but the cause(s) are unknown'. In order to demonstrate how difficult it is to distinguish the effects of known agents from 'recent forest decline' in just one species, Table 7.5 lists a number of the commoner agents with their distinctive symptoms that affect Norway spruce which can be superficially confused with the symptoms of 'recent forest decline'.

Dendrochronology (i.e. the use of tree-ring data to provide information on past performance of trees) has been used in many studies to follow the progress of problems associated with air pollution. One of the best, from which the results shown in Fig. 7.3 are taken, has been done in the Rhône valley where there are sources of both sulphur dioxide and fluoride. By combining observations of the width

Long-term fluoride pollution of a forest ecosystem

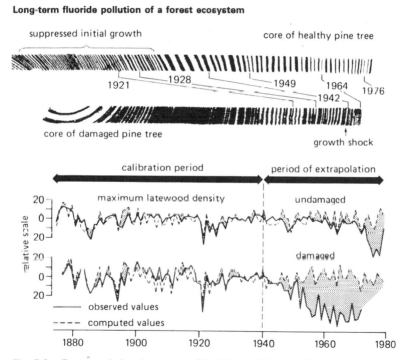

Fig. 7.3 Cores and densitograms of healthy and damaged pine trees in the lower Rhône Valley. In this particular case the reduction of growth in the damaged cores could be directly related to the introduction of a second aluminium smelter around 1947 emitting sulphur dioxide and fluorides. (Courtesy of Dr H. Flühler, Birmensdorf, Switzerland and D. Reidel Pub. Co., Dordrecht, Netherlands.)

Table 7.5 Known causes of damage to Norway spruce (*Picea abies*) which can be confused with 'recent forest decline' damage

Agent	Symptom
Abiotic	
Low temperatures or 'winter chlorosis'	Needles of all ages slightly yellow in early spring
Drought or salting of nearby roads	Red coloration of older needles on branches in exposed localities
Pathogenic	
Spruce needle cast fungus (*Lophodermium piceae*)	Older needles red-brown with black spots or bands in spring
Spruce needle scorch (*Lophodermium macrosporum*)	Black spots along central rib of needle
Rizosphaera needle scorch (*Rhizosphaera kalkhoffii*)	Like the above symptoms but with much smaller black spots
Spruce needle rust (*Chrysomyya abietis*)	Orange banding of needles
Grey mould (*Botrytis cinerea*)	New needles go brown and hang down; isolated patches of infection; not to be confused with frost damage
Sawfly (*Paristiphora abietina*)	Young needles eaten on one side only, rest turns red
Spruce bell moth (*Epinotia tedella*)	Leaf bases eaten and needles turn red-brown
Spruce bark beetle (*Ips typographus*)	Bark peeling and resin droplets on trunk, trees ultimately become red-brown and die

of tree rings with a computer analysis, which takes into account monthly rainfall and temperature, this study was able to show that the start of the damage may be traced back precisely to the time when another aluminium smelter was introduced into the area and that climate, in this case, was not the prime cause of growth retardation. Other less precise dendrochronological studies of 'recent forest decline' have revealed similar patterns of reduced tree ring growth dating from the late 1960s and early 1970s which also indicate that fluctuations in climatic conditions alone are not responsible. These less-accurate studies have generally shown that 'recent forest decline' affected tree ring growth at higher altitudes first and the occurrence of these symptoms in other trees gradually spread down the slopes of the afforested regions. Nevertheless, tree-ring studies are ambivalent as to cause, except in certain cases (e.g. Fig. 7.3), because it is impossible to isolate one factor from the many which collectively might be responsible for the reductions in tree growth.

(b) Possible causes

Because the exact causes of 'recent forest decline' are not known at present, much effort has been devoted to the identification of possible ones, and dispute has surrounded those hypotheses put forward as explanations in the scientific literature and by the environmental media or lobbies. Table 7.6 lists a number of likely hypotheses which have been put forward as explanations of 'recent forest decline' roughly in the order in which they have been propounded, although there are many others (over 150 at the last count). Explanations based on numbers 2, 3 and 4 in the list are currently thought to be amongst the most likely.

(c) The 'bad-practice' hypothesis

It is a long held view of certain foresters dating back to the turn of this century that there were many inherent dangers in establishing large new forests (or replanting old ones) with a single species of conifer, often of the same provenance, especially if it was naturally non-native to the region of planting. Such monocultures were held by these prophets to be ultimately more prone to attack by natural pathogens and when this happened the destruction would be on a larger scale than usual especially as the trees got older and reached maturity. Moreover, because of extensive nature of these plantings, areas with non-ideal soil conditions would have to be included which would then ultimately be revealed as unsatisfactory especially if non-native vegetation was planted.

211

Table 7.6 Hypotheses suggested as possible causes of 'recent forest decline'

Hypothesis name	General features†
1. 'Bad practice'	Bad forestry practices including use of foreign species and monoculture on unsuitable soils
2. 'Acid rain' or 'Soil leaching'	Acidic (mainly wet) deposition causes soil leaching of essential minerals and increases the availability of toxic ions such as aluminium to root and mycorrhizal associations
3. 'Ozone' or 'Photochemical'	Action of sunlight causes enhanced production of ozone, PANs and hydrogen peroxide, etc., which promote free-radical attack on vegetation
4. 'Modifier' or 'Pollution-enhanced stress/infection'	A series of droughts or unusually cold winters or 'enhanced stress/infection', persistent attacks by pathogens (including viruses), along with general atmospheric pollution, weakens still further aged, sensitive and exposed trees
5. 'Ammonium' or 'Excess nitrogen deposition'	Oversaturation of trees with nitrogen which grow too fast and are thus easily attacked by parasites or rendered more prone to stresses
6. 'Chloroethane' or 'Photoactivation'	Presence of halocarbons initiates ultraviolet light-initiated damage in photosynthetic mechanisms
7. Alternatives (many)	Wide variety of possibilities including tetra-alkyl lead from car exhausts exerting toxicity when degraded to trialkyl lead; dinitrophenols, PCBs and aromatics also phytotoxic or latent viral infections cause epidemics
8. Multi-interaction	Any of the above causes combined together as equals or as one primary and one (or more) secondary cause(s)

† Much abbreviated, see text for details.

These 'I-told-you-so' predictions are not very helpful, 60 years or so later, as an answer to the cause of 'recent forest decline'. Perhaps their use lies in any fresh consideration of future Planting strategies. On the other hand, it may simply be that some forest trees have been allowed to grow too long before felling and replacement. Many of the affected trees, are now well over 60 years old and hence more prone to stress anyway.

(d) The 'acid rain' or 'soil leaching' hypothesis

This theory, largely attributed to Ulrich, suggests that the main effects are brought about by excessive acidity reaching forest systems which brings about changes in soil chemistry in a succession of stages. Initially, extra inputs of sulphur and nitrogen cause an increase in tree growth by virtue of their fertilising effects but, in the longer term, acid deposition reduces the neutralising effect of the soil and ultimately exceeds the buffering capacity of the carbonic and silicic acids (see Ch. 4). The consequences are many. The ability of the soil to continue to provide essential nutrients such as calcium, potassium and magnesium for tree growth will be diminished because these will be leached from the soil around the roots (as well as from the needles or leaves) and eventually lost in run-off groundwaters. Losses of magnesium, and possibly manganese, could give rise at least partly to the yellowing of certain needles. In addition, aluminium ions will be mobilised from soil particles into solution (Sect. 4.2a) and become toxic both to uptake mechanisms on the fine root hairs and to their partner fungi in mycorrhizal associations with the roots, thereby diminishing their ability to take up essential minerals and water. This then leads to a destabilisation period when trees become more susceptible to stress caused by disease or drought but, more than that, increased needle loss, decay and other processes involved in this complex ecosystem trigger off other acid-forming mechanisms which make the problem worse.

There is considerable evidence to support this hypothesis. Many studies have shown higher levels of protons and aluminium ions and reduced levels of calcium, potassium, magnesium and manganese when simulated acid rain has been applied to representative forest soils where 'recent forest decline' has occurred. Certain forest soils in southern Germany have been shown to become magnesium-deficient and, as a result, the trees that grow there have greater amounts of yellow second-year needles which have abnormally low magnesium contents. They may be distinguished from yellowing caused by fungal attack because they remain on the trees whereas after fungal infection they fall off.

Spruce seedlings are relatively tolerant of high levels of acid and low levels of calcium, unlike beech seedlings, but are sensitive to aluminium ions. However, root damage caused by aluminium does not occur if the pH is raised above pH 4.5 when precipitation of aluminium hydroxide takes place, or if extra calcium is added, or if the aluminium is supplied in organic form.

At the moment, this highly plausible hypothesis in its original form remains unproven because a number of other investigative studies have been unable to find a clear relationship between acid deposition, aluminium mobilisation and 'recent forest decline'. It could, however, be a partial explanation when combined with some of the other possibilities, especially when magnesium-deficiency is a possibility on certain soils.

(e)   The 'ozone' or 'photochemical' hypothesis

This hypothesis suggests that the needle (or leaf) damage associated with 'recent forest decline' is caused initially by higher levels of ozone (see Ch. 5) which do not decay away quite as fast at higher altitudes as they do during more usual diurnal cycles (see Fig. 5.3) at sea-level. Increased levels of hydrogen peroxide and other secondary photo-oxidants have also been suggested as having the potential to cause similar needle (and leaf) damage. They have been included here as an optional extension to this possible mechanism rather than give them separate status as has been claimed by some.

Separate effects of ozone and other oxidants upon vegetation have already been covered in Ch. 5 but this hypothesis goes on to suggest that, in the case of 'recent forest decline', the oxidants damage cell membranes and allow the leaching of nutrients (organic and inorganic) from the needles or leaves which then allows subsequent attack by pathogens, like insects and fungi, or other symptoms of stress to develop. On first reflection, such a hypothesis might explain rather better than others the enhanced yellowing of needles sometimes associated with 'recent forest decline' and the 'damage-uppermost' symptom often observed in the secondary branches as being due to direct attack of photochemical agents. However, yellowing of needles also occurs during magnesium deficiency but, away from proven magnesium-deficient soils, it is not known if this arises as a result of the leaching caused by photochemical attack or as a result of a failure by the roots to take up sufficient magnesium.

One interesting adjunct to this hypothesis is some explanation of a partial melting on affected trees of the thin 'waxy plugs' which are normally stretched across the sunken stomata of young conifer needles. If ozone moving inwards should meet ethylene and hydrocarbons moving outwards from needles then free radicals, such as hydroxyl radicals, would be formed in close vicinity to the stomata which would then immediately attack and degrade the nearby waxes of the cuticles, thereby changing the normal fluxes of carbon dioxide and water across these thin coverings. Cracks in the thin waxy plugs (as well as 'melting') have been observed as a result of surveys of conifer needle samples examined by scanning electron microscopy. It has been claimed by some that these cracks are caused by acid rain rather than by ozone.

Studies to confirm or refute the 'ozone' (or 'photochemical') hypothesis are conflicting. Experimental fumigations with just ozone have usually failed to produce the visible symptoms characteristically associated with 'recent forest decline'. This has been especially the case with Norway spruce, *Picea abies*, which is classed as ozone-resistant in the USA. Nevertheless, reductions in the rates of photosynthesis in silver fir, *Abies alba*, have been detected during long-term fumigations with ozone. Scots pine, *Pinus sylvestris*, however, is ozone-sensitive and fumigations have caused chlorotic mottling of the needles as well as reduced root growth and have increased the rate of senescence of older needles. Structural damage to certain needle cells having high rates of transport (i.e. the Strasburger or albuminous cells) characteristic of ozone attack has also been detected in the needles of experimentally fumigated and field-affected conifers.

The 'ozone' (or 'photochemical') hypothesis is a plausible explanation but more (and better) experimentation is still required to verify or refute it as a primary cause by itself. Almost all the existing evidence is circumstantial and it is likely that the relative importance of ozone is different in different areas and upon different species. Ozone may well turn out to be a primary cause of forest damage in North America and only a contributory factor to 'recent forest decline' in Europe.

(f) The 'modifier' or 'pollution-enhanced stress/infection' hypothesis

Sulphur dioxide by itself in gaseous form has, by and large, been discounted as a primary cause of 'recent forest decline'. In fact, levels of sulphur dioxide in affected areas have been steady for some time or even have been falling recently. Neither fumigation studies nor surveys in areas around local sources of high emissions of sulphur dioxide have revealed close similarities with the new phenomenon. Moreover, the depleted branches of affected trees tend to be covered with lichens which are often sensitive to sulphur dioxide although, normally, the particular species of lichens involved are those more tolerant of acidic conditions. Similarly, there is little evidence that oxides of nitrogen alone could cause such symptoms directly although they may contribute excess nitrogen (see next section).

The case for an interaction between sulphur dioxide and oxides of nitrogen, especially with intermittent exposures to ozone, is much better. As Sect. 7.1(c) has already outlined, these mixtures are known to be capable of causing more-than-additive reductions in growth and an accelerated appearance of injury in a variety of plants. Such mixed exposures could be causing the needles or leaves in exposed, aged or sensitive trees to be more readily attacked by pathogens or more predisposed to stress injury. A full range of possible stresses has already been outlined (Sect. 7.1a) but, in the regions affected by 'recent forest decline', some of the most important and different stresses (which would include either frost injury or winter desiccation, drought and wind) are very much interrelated one to another. For example, wind increases transpiration which is undesirable during drought and low temperatures render water unavailable for use because it is frozen. It is known that the rates of water uptake by roots are greater in those plants that are subjected to mixtures of atmospheric pollution and that this is made worse because similar combinations decrease the relative proportion of roots (by mass) and increase the relative amounts of the affected plant above ground (i.e. the root/shoot ratio falls). This means that the reduced amount of root tissues have to work much harder to sustain a bigger proportion of above-ground tissue during droughts, etc., and so the susceptibility to damage caused by losses of water will be greater. This is a very real possibility in forests with soils that do not retain large amounts of water. In conifers it is also known that losses of water due to atmospheric pollution occurs across the cuticles rather than through the stomata. Structural or physical changes caused by combinations of pollutants to all the waxy coverings rather than just the melting or cracking of the waxy plugs of the stomata (by ozone and acid mists, respectively) could therefore be critical.

Climatic fluctuations over the last century are not sufficient in themselves to explain the increased incidence of 'recent forest decline' due to stresses such as cold, drought or wind alone although it has probably been exacerbated by a few dry summers followed by unusually cold winters (often with false springs) more recently. A question provoked by this hypothesis is: 'Does the adverse cold, drought or wind stress (or sustained infection) predispose the trees to damage by the mixture of atmospheric pollutants or has their additional presence rendered them more susceptible to subsequent stresses or infections?'. The answer is not known but experiments are still in progress to evaluate this question. It is more than a suspicion that the answer will be that it does not matter which comes first, one set will predispose the trees to the next.

(g) The 'ammonium' or 'excess nitrogen deposition' hypothesis

One of the most evident and recent changes in the patterns of air pollution emission has been a shift in the balance from sulphur dioxide towards nitrogen-based air pollutants as a result of increased (mainly mobile) combustion, increased applications of artificial fertilisers, sewage treatment, and more intensive animal husbandry (see Ch. 3). Very high levels of atmospheric ammonia are now being recorded in the Netherlands and surrounding countries which may have caused enhanced incidence of ozone injury (see above) or an increased deposition of nitrogen (in both wet and dry forms) on to sensitive ecosystems such as forests. This could become the start of fresh problems because there are good reasons to believe that a number of ecosystems are harmed by inputs of extra nitrogen.

On the basis of some of these appreciations, Nihlgård and others have put forward a wide-ranging hypothesis which may account for 'recent forest decline'. They postulate that, as a result of higher rates of nitrogen deposition into forests, there is initially an excess of growth which leads to an accumulation of toxic waste products because the balance between non-nitrogen- (i.e. carbohydrates) and nitrogen-containing metabolites (proteins, etc.) is disturbed and valuable extra minerals (phosphate, magnesium and potassium) are consumed as the plant attempts to re-establish a balance, thereby leading to internal deficiencies. The toxic products (amides, amines and ammonium derivatives) eventually cause an increase in the rate of cell death or, after exudation, encourage fungal attack or growth of epiphitic leaf cover (i.e. increased lichen and algal growth) and leaching, which, in turn, reduces photosynthesis. At the same time, the roots are deprived of carbohydrates, because these are needed above to compensate for the extra nitrogen, which means less carbohydrates are available for the vital mycorrhizal associations of roots with specific soil fungi. This reduces still further the amount of microbial activity in the soil which is already depressed because the extra nitrogen coming into the soil by excess nitrogen deposition reduces the necessity for nitrogen fixation. Such general changes in balance caused by reduced root and enhanced leaf growth may then render forest trees more susceptible to other stresses imposed by drought or wind-blow or to infection or attack by pathogens attracted by the increased leaf exudations.

At the moment, there is little solid experimental evidence to substantiate this hypothesis in relation to 'recent forest decline' but many indications of the existence of the separate components have been observed. Nitrogen-based atmospheric pollutants contribute

Table 7.7 Ecosystems thought to be especially sensitive to excessive inputs of nitrogen

Type	Details
Wetlands	1. Ombrotrophic mires (i.e. bog-like areas only supplied with nutrients from rainfall), e.g. raised bogs 2. Mires in granitic areas or otherwise nitrogen-limited
Lakes	1. Clear water lakes partially covered with certain species (e.g. water lobelia, quillwort and shoreweed) 2. Nutrient-lacking lakes with a high number of pondweeds including *Potamegeton*
Others	1. Heathlands with high amount of lichen cover with limited mineral availability 2. Meadows with limited mineral availability used for extensive grazing and haymaking to which artificial fertilisers have never been added 3. High-altitude coniferous forests

around 30 % of the acidic wet precipitation. After heavy snowfall and snow melt, the contribution of nitrate can be particularly high and this often coincides with the most sensitive stages of plant development. Moreover, alterations to growth and increased susceptibility of forest trees to different stresses (Table 7.1) may be just one example of a broader problem of extra nitrogen which may be more widespread and not just confined to forest trees.

Ecosystems thought to be most sensitive to nitrogen-based atmospheric compounds are those which contain certain species of plants which are specially adapted to low levels of nitrogen and nutrients. Those ecosystems in this category and which are considered most threatened by enrichment with nitrogen include those listed in Table 7.7. Effects observed in such systems are mainly due to increased competition from faster-growing plant species which are normally restricted in these ecosystems because of the low availability of nitrogen. Ultimately, this may lead to the disappearance of the original plant species characteristic of such ecosystems, with a loss, ultimately of irreplaceable genetic resource.

For example, in bogs only supplied with nutrients from rainfall (ombrotrophic mires), certain mosses (*Sphagnum*) are replaced by seed plants. Similarly, heathlands show a move towards heather-like plants whilst meadow vegetation is threatened by a spread of nitrogen-demanding tall perennials like thistles, nettles or rosebay willow herb and lakes grow more plankton. Vegetational changes can also exert changes upon animal populations. For example, fish may be starved of oxygen because of competition with rapidly growing plants or explosive attacks of bark beetles may occur in enriched forest stands.

(h)   The 'chloroethene' or 'photoactivation' hypothesis

The novel nature of 'recent forest decline' has lead Frank to suggest that freshly introduced atmospheric pollutants may be the prime culprits. Such new chemical candidates must have half-lives in the troposphere sufficiently long to be transported to rural areas at higher altitudes and to be toxic to plants at relatively low concentrations. His research team examined a wide range of halocarbons that are increasingly used as refrigerants, etc. (see Sects. 5.1b and 6.2a) and they concluded that the chloroethenes (trichloroethene and tetrachloroethene) are indeed capable of causing phytotoxic effects similar to those associated with 'recent forest decline'. This damage was especially severe so in the presence of strong light enriched in the ultraviolet region around 280–320 nm. Such radiation is to be found only at higher altitudes because it is strongly attenuated by misty, lower urban and semi-urban atmospheres.

Toxic effects first start in the lipid layers of plant cells, where the chloroethenes naturally accumulate and the ultraviolet light causes their conversion into free radicals which have been claimed to degenerate to form, for example, highly reactive atomic chlorine, the phytotoxic dichloro-acetylene and even phosgene. The photoactivated destruction that has been attributed to the chloroethenes is primarily concentrated upon the thylakoid membranes of chloroplasts packed with highly sensitive pigments vital to photosynthesis, such as chlorophyll a and β-carotene (see Appendix, Sect. 8.3). This may have produced the characteristic bleaching and yellowing symptoms so often associated with 'recent forest decline'.

As yet, there have been no comprehensive studies to evaluate chloroethene-induced injury as a primary cause of 'recent forest decline'. Some authorities have dismissed this possibility by saying the levels are far too low but this misses the point that most halocarbons are extremely persistent and, moreover, alien to the environment. There are no organisms to degrade them and surveys have shown them to be rising steadily in clean-air regions from virtually nothing in 1970 to around 0.37 ppbv (2.5 μg m^3) now. Furthermore, they are likely to be concentrated by a number of processes. Calculations have shown, for example, that tetrachloroethene in gaseous form around a leaf may attain a two-thousandfold increase in concentration in the lipid or waxy phases of a leaf. This particular hypothesis may warrant more attention than it has hitherto drawn from other researchers largely because we are all unfamiliar with many of the hazards presented by halocarbons in the atmosphere.

(i) Alternative hypotheses

There are many other hypotheses which have been put forward as possible explanations of 'recent forest decline'. These break down into two different types. There are those like the chloroethene hypothesis which claim that organic micropollutants of various types might be the cause. From among the 1500 or so organic compounds, that have been identified in air samples gathered over semi-urban and rural areas originating as man-made emissions, there are a number which could be harmful. For example, a recent report from West Germany has indicated that, there are occasionally episodes when the levels of tri-alkyl lead may be a thousandfold higher in rainwater samples than was previously thought to be possible. Tri-alkyl lead is a degradation product produced from tetra-alkyl lead added as an anti-knock agent to gasolene and, at these higher levels in the atmosphere (0.3 μM), they could be phytotoxic. Other micro-organic pollutant claims have been made for the dinitrophenols, which are potent inhibitors of respiration and mineral uptake, and for a variety of other groups or organics such as the polyaromatic hydrocarbons (naphthalene, anthracene, indene, benzpyrene, etc.) which might be partially oxidised before possibly exerting their phytotoxic effect. Similar hazards to forest trees have been attributed to the persistence of pesticides and herbicides (e.g. polychlorinated biphenyls, or PCBs).

The alternative to organic micropollutants is that a biological infection is responsible for initiating 'recent forest decline'. Within those suggestions, which might be termed collectively the 'epidemic' hypothesis, an improperly understood virus and its vector could be involved in propagating a disease throughout forest trees. This means the virus particles may have been spread much earlier and reside in a dormant phase for some time before conditions are ripe for the active phases of the disease. There are many examples of widely spread and apparently innocuous viral particles occurring throughout the plant kingdom. The *Potex* type of virus, for example, has been found in reasonable proportions (2–22 %) in all those tree species known to be affected by 'recent forest decline'. The nature and mode of spreading of any putative viral infection is not known but is presumed to be by insects, needle spores, root myccorhizal hypae or animals in the soil. At present, because the nature and spread of 'recent forest decline' does not fit well with most of the characteristics shown by other types of viral infection in plants, this hypothesis is not currently viewed with much favour.

(j) The 'multi-interaction' hypothesis

It is more than likely that 'recent forest decline' is one more facet of the problems that arise from the activities of modern society which have become more complex and increasingly more threatening to the natural environment. Each of the hypotheses covered in the previous sub-sections has its merits and most of them imply interactions as primary or secondary causes. It is difficult for both the experienced researcher and the interested reader to make a judgement on the varied merits or relative importance that should be ascribed to each of them without more definitive experimentation. Even if critical comparisons were to be undertaken, the problem of distinguishing the significant from the 'background noise' still remains. If it were, hypothetically, possible to accomplish this the suspicion is ever present that the answer will never come in a clear-cut manner favouring one hypothesis over all the others but most likely as an interacting combination of many of the components of the various hypotheses already described. Alternatively, it could be something entirely new and unsuspected hitherto which is the primary cause which is then followed by secondary interactions of the types already mentioned. To cover these possibilities, it would be wise at the moment to allow some degree of compromise by means of a multi-interaction hypothesis.

Evidence for one possibility along these lines has recently been obtained in my own laboratory. We have found that a short ozone fumigation (150 ppbv for 7 hours) causes immense visible leaf injury to pea seedlings the day after treatment. If, however, the seedlings are grown up in ozone atmospheres they are quite undamaged. This tolerance is related to the fact that these preconditioned seedlings produce only low amounts of ethylene (a natural plant growth substance which in trace amounts may interact with other regulators to coordinate a variety of developmental responses) whereas those injured by a burst of ozone produce large quantities of 'stress' ethylene. These emissions of extra ethylene can be prevented by treating the plants with an inhibitor (aminoethyoxyvinylglycine) before the ozone fumigation and this then protects the plants and they then do not show the characteristic visible injury caused by a sudden exposure to ozone.

Sudden exposure of similar seedlings to oxides of nitrogen (instead of ozone) also causes the production of similar amounts of 'stress' ethylene to that caused by a sudden exposure to ozone but this also does not cause visible injury. This means that ozone and ethylene must be reacting either in the plant cells or in the air spaces within the leaves to generate harmful free radicals which then cause extensive peroxidation (see Ch. 5) and this leads to visible leaf injury.

Short fumigations with mixtures of ozone and oxides of nitrogen, however, do cause the evolution of ethylene and visible leaf injury.

Our results therefore indicate that intermittent episodes of ozone pollution are more phytotoxic than continuous high levels of ozone and there is greater tolerance to ozone when rates of ethylene formation are depressed. The additional presence of other pollutants such as the oxides of nitrogen and additional environmental stresses such as mineral deficiencies, chilling, drought, etc., may cause an increase in ethylene evolution which then enhances peroxidation and leaf injury. This type of explanation therefore combines certain aspects of several of the hypotheses already mentioned and could be a likely step in providing clues to solve a complex problem.

7.3 GENETIC ADAPTATION IN RESPONSE TO AIR POLLUTION

(a) Industrial melanism

Evolution occurs more rapidly when organisms are exposed to fresh stresses and, in some cases, this can be quite a quick process. For example, the ability of certain bacteria to circumvent antibiotics and other drugs is a major cause for concern in the medical profession. The speed at which they can do this is often faster than the ability of the drug companies to introduce fresh and effective drugs onto the market. By similar mechanisms, certain insect populations acquire immunity to insecticides, rats to rodenticides, and plants to herbicides, although it is fortunate that they do this at a slower rate than bacteria.

The classic case of atmospheric pollution darkening the barks of trees, which then affects the colour of moths, is often quoted as an example of genetic adaptation but it is worth a more detailed consideration because it illustrates important factors and may correct some misconceptions. The whole phenomenon is called 'industrial melanism' and was observed in first one species of Lepidoptera (moths and butterflies), then others, and new forms of 'industrial melanism' are still being detected. The overall process is relatively fast but starts slowly, enters an exponential phase, before slowing again. The frequency of dark (melanic) individuals in the population thus shows a sigmoidal relationship with time. The first detected occurrence of industrial melanism was in 1848 when a black specimen of the common peppered moth, *Biston betularia*, was found in Manchester and was called the *carbonaria* form (Fig. 7.4a and b). By 1895, over 95 % of the individuals in the population around Manchester were black though there is only one generation a year. In the UK, there are now over eighty different species recorded as being affected by industrial melanism and many other examples now exist elsewhere in the world although the bulk of them have one feature in common. Most of them rest with their wings fully open on tree trunks or rocks and derive protection from predators (mainly birds) by having colorations and patterns on their upper wings which closely resemble the background on which they rest. Many species also have blacker larval stages as well but this is controlled by a different set of genes.

It was actually quite a long time before a full explanation of the mechanism of industrial melanism was available because entomologists and ornithologists did not often collaborate. Studies by Kettlewell in the 1950s showed that escape from predation by birds is a powerful agent of natural selection because a large number of those birds which hunt by sight actively search tree trunks for resting insects. Insects, on the other hand, do not alight on those trees which suit their colouring although they will move around once landed to accord better with their surroundings. There is thus a balance of wing (or larval) coloration of the population as a whole which is determined by the feeding activities of the birds in a particular environment. As the surroundings become blackened by deposition of particulates and enhanced sulphide formation on surfaces, the total gene pool (a collective expression for all genetic characters) of the population moves in favour of wing (or larval) blackening to correspond more closely with the surrounding environment in areas of atmospheric pollution. Whether this evolution takes place due to selective pressure by means of favouring small genetic variations already within a population or by the chance appearance of favourable mutants involving a sudden genetic change is still one of the big arguments of evolutionary theory. In every species showing industrial melanism, however, the trait is characterised by the spread of a dominant or semi-dominant mutation. If it were recessive then the offspring would not be melanic and any advantage would disappear. The rapid response to darker surroundings by virtue of adaptation therefore lies in the accumulation of genes that are at least partially dominant in their effect.

The original *carbonaria* specimen of the common peppered moth was a heterozygote (i.e. it carried genes for both melanic and normal colouring) and was actually much lighter in colour than both the heterozygotic and homozygotic 'modern' *carbonaria* forms that were trapped much later and, indeed, up to the present time. Crosses of 'modern' heterozygotes with 'normal' moths from non-industrial localities will still produce lighter-coloured 'original' *carbonaria* forms. In other words, industrial melanism is an excellent example of the evolution of dominance whereby the advantageous melanic mutant trait has been strengthened by the presence of other genes which converted the partial dominance of the 'original' forms into the

(a)

Fig. 7.4(a) and (b)   Specimens of the common peppered moth (*Biston betularia*) showing typical and *carbonaria* forms resting upon a lichen-covered tree in rural Dorset, UK (a) or a blackened lichen-tree in industrial Birmingham, UK (b). Note: There are two specimens in both pictures! (Courtesy of Professor E. B. Ford, FRS, Oxford, and Methuen & Co., London.)

complete dominance of the 'modern' *carbonaria* type. It is thus the whole gene pool within a population that forms the basis of successful adaptation.

Single gene mutations are, by and large, unable to respond to a particular environmental stress and it is the formation of higher proportions of active gene combinations which gradually increases the effectiveness of a population to resist such a stress. Certain individuals within a population may be particularly well endowed with these useful combinations and, as such, worth further selection by breeding programmes. In the case of tolerance to air-pollution stresses, a

(b)

selection by man of animals or plants on this basis has rarely been made although there have been attempts to improve the tolerance of tobacco and eastern white pine to ozone. There must be considerable advantage in searching for (and exploiting) similar tolerances or adaptations to a wider range of pollutants in other species in the future. Despite the vast sums that are spent on fresh abatement measures, atmospheric pollution is likely to be present for the next century and is likely to increase especially in parts of the world where it did not occur before. This means that certain ecosystems will have only a short time to build up natural tolerance to stresses induced by air pollution.

(b)   Sensitivity and tolerance

Evolution of resistance by plants is a well known phenomenon. By such means plants may survive and grow in the presence of high concentrations of heavy metals or salts, in conditions of prolonged drought, or in a wide variety of seemingly adverse conditions. Resistance to pollution by plants has been defined by Bradshaw as 'the ability to maintain growth and remain free from injury in a polluted environment'. It is a relative phenomenon whereby resistance need not be complete and takes into account two possibilities: stress avoidance, whereby pollutants are excluded, and stress tolerance (i.e. the possession of a pool of genetic characteristics that confer an improved ability to detoxify, repair or compensate for injury caused by atmospheric pollutants) when detoxification mechanisms exist to counteract incoming pollution. In the case of resistance to the deposition of atmospheric pollutants, the avoidance component does not apply because plants have to open their stomatal apertures to function and, consequently, the mechanisms involved are those of stress tolerance and not those of stress avoidance.

Perhaps the best practical illustrations of tolerance to atmospheric stresses have been shown by grasses although it has been also demonstrated in geraniums, French beans, soyabeans and several tree species. The dairy farmers on the higher ground around Manchester in the UK, and probably those downwind from many other industrialised cities, have long known that any attempt at re-seeding of their pastures with commercial grass selections would be met with a very poor germination and, even if successful, stunted growth for some time afterwards. Only when sufficient, previously adapted strains of grass have had time to spread in from the surrounding pastures or take over as a result of germination of buried seed or regrowth of older root fragments which gradually out-compete with the resown grasses, could the previous conditions be re-established. These farmers, and most likely their fathers and grandfathers before them, quickly realised that their existing grass swards were very much better suited to growth in their particular areas. They may have also appreciated that atmospheric pollution was somehow involved but plant breeders have never taken much notice of the phenomenon. Perhaps this is because hardy hill grass stock is involved and not lush grass varieties growing in optimum conditions or cereal cultivars with high demands on artificial fertilisers.

The pool of genetic characters in these particular grasses and in other vegetation from similar surroundings must be of value especially if they can be combined with other desirable growth characteristics. The hills around Manchester have been experiencing atmospheric pollution for over 200 years longer than anywhere else in the world. The adaptations to the gene pool of these particular grasses, that have undoubtedly arisen over this period because of atmospheric pollution, could be extremely useful in other polluted localities.

Even highly selected cultivars of grasses and cereals possess a wide variation in sensitivity or tolerance to single or combined atmospheric pollutants. Recent work by some of my colleagues at the University of Lancaster illustrates this point extremely well. They carried out a very simple experiment (the final results of which are shown in Fig. 7.5) in which they sowed grass seeds of a popular variety of smooth meadow grass, *Poa pratensis* cv. Monopoly, currently recommended for inclusion in grass re-seeding mixtures in the UK, and grew the germinated seedlings right through to seed head formation in clean air, or in air polluted with low levels (68 ppbv) of sulphur dioxide or nitrogen dioxide, or a mixture of the two. At harvest (i.e. just after when haymaking would normally take place), they ranked the seedlings according to growth and seed-head formation as shown in Fig. 7.5. As may be clearly observed, some individuals from the single-polluted, and even the combined-polluted, treatments are as good, or even better, than some of the best of the individuals grown in 'clean air'. This clearly means that, even though this variety is highly selected on the basis of the 'desirable' characters chosen by plant breeders (growth characteristics, digestibility, total nitrogen, etc.), there is still an untouched reservoir of genetic variation towards atmospheric pollution stresses. Furthermore, if this is the case for one variety of one grass, then there is a strong likelihood of a similar situation existing for a wide number of other species.

One of the most frequent limitations to plant growth after light, temperature and water is an adequate supply of nitrogen. Tolerance and sensitivity to oxides of nitrogen within a plant population has never been demonstrated (apart from the simple experiment shown above) merely because it has never been looked for. Agricultural and horticultural plants that can tolerate oxides of nitrogen with additional amounts of sulphur dioxide, or even use the atmospheric oxides of nitrogen as a source of nitrogen, will become more valuable as time passes.

During glasshouse cultivation (Sects. 3.2d and 6.2b), some of the highest atmospheric levels of oxides of nitrogen are encountered by plants (and growers). Recent studies have indicated that certain modern tomato cultivars used in carbon-dioxide-enriched glasshouses perform better than older, commercial cultivars especially when oxides of nitrogen are also present. This means that the plant breeders of tomato cultivars, who use commercial glasshouses as part of their selection process, have actually followed the traits of tolerance towards oxides of nitrogen and increased the genetic pool of protective characteristics within their more recent products without realising that they were doing so. Similarly, in the experiment shown in Fig. 7.5 there are individuals within the nitrogen dioxide treatment and also the nitrogen dioxide plus sulphur dioxide treatment that perform as well, if not better, than the majority of the 'clean-air' grown plants.

Fig. 7.5 Ranked individual plants of smooth meadow grass (*Poa pratensis,* var. Monopoly) grown to harvest in, A, clean air, B, NO_2-polluted air (68 ppvb), C, SO_2-polluted air (68 ppvb) and, D, SO_2 plus NO_2 (68 ppvb each) polluted air for 20 weeks. (Courtesy of Dr Whitmore and Prof. Mansfield FRS., Lancaster.)

Chapter 3 has described evidence which suggests that plants take up oxides of nitrogen and use them within their own metabolism of nitrogen partly for nutrition and partly for detoxification. At the moment, the amounts of nitrogen involved are but a small fraction of the total nitrogen demands of a plant even though nitrogen dioxide-treated plants are greener, and sometimes 'leafier', than unpolluted plants. Nevertheless, the fact that they can handle these pollutants and acquire tolerance to atmospheric pollution stresses means that such processes are taking place naturally within those ecosystems that can tolerate extra nitrogen, unlike the sensitive ecosystems described earlier in Sect. 7.2(h). This means that plant scientists and breeders could accelerate a selection process towards this end without resorting to complex genetic engineering and produce new cultivars which require less artificial fertiliser. Unfortunately, this would be at a 'cost' because tolerance or detoxification mechanisms are introduced at the expense of overall yield. For agricultural crops, it all depends on how these 'costs' are evaluated; whether they are based on sound socio–agronomic–economic principles or if they are at the whim of politicians. For other species in the environment, natural selection is the arbiter.

7.4 OVERVIEW

(a) Models

One of the most overworked words in the English language is the word 'model'. Chemists and pharmacologists use model compounds to elucidate general mechanisms and computer scientists or statisticians develop software models which respond to different inputs. Other scientists, particularly biologists, study model systems (i.e. a particular life cycle or ecosystem) to develop ideas and overall concepts. When these are written down it often helps to have a conceptual diagram of the different processes, some of which may be expressed mathematically, if sufficiently defined, or at least as a flow diagram using arrows to indicate interdependencies. These paper models help others to appreciate what is important, or what is trivial, and indicate what might be interesting to investigate next. They may be viewed as a series of hypotheses supported by observation and theory which could be developed still further, even to the extent of being expressed in terms of mathematical or chemical formulae, but are a help to the understanding of a complex problem.

The topic of atmospheric pollution effects on living systems is certainly one of the most complex to face environmental scientists. We only know less than 10 % of what concerns living systems and, similarly, less than 10 % about all possible air pollution effects. This means that on the basis of a small fraction, projections are made, implications assessed and appropriate action advised. Nevertheless, this has to be done; it would be foolish to delay. The problems stemming from atmospheric pollution are already with us, and some of them are of major global concern.

(b) Vegetational impact

There have been a number of paper models to illustrate the effects of different pollutants upon different plant systems. Some of the best examples have been produced by Ulrich to illustrate the 'acid rain'/'soil leaching' hypothesis concerned with 'recent forest decline'. However, none of them attempt an overview where changes external to a plant in the context of atmospheric pollution are simultaneously considered in the relation to possible internal responses. One attempt is shown in Fig. 7.6. The following paragraphs endeavour to lead the reader to an understanding of our model (despite its apparent complexity) by using alphabetical identifiers for each stage on the diagram and in the text.

Man-induced emissions of sulphur dioxide and oxides of nitrogen increase atmospheric concentrations and the action of sunlight causes the formation of secondary pollutants like ozone (A). In time (hours to months) these pollutants are transferred to terrestrial and aquatic ecosystems by either dry- or wet-deposition mechanisms. Rates of dry deposition (B) vary widely but they are of similar magnitudes for all the major atmospheric pollutants. Meanwhile, a complex series of oxidations (mainly in the wet phase of raindrops, etc.) increase the amount of acidification (C) within the atmosphere. The amount of injury caused directly by wet deposition (D) on vegetation is still unresolved but surface oxidation within the liquid film continues and penetration of the plant cuticles (H) may cause leaching of critical ions and modification of the rain falling through tree branches on to the soil (E). Acidic deposition leaches calcium and magnesium from soils and causes the mobilisation of aluminium ions which damage tree roots and influence the uptake of both nutrients and water F.

Over 80 % of the total pollutant flux during dry deposition enters the stomata (G) and passes into the cells across a variety of phases

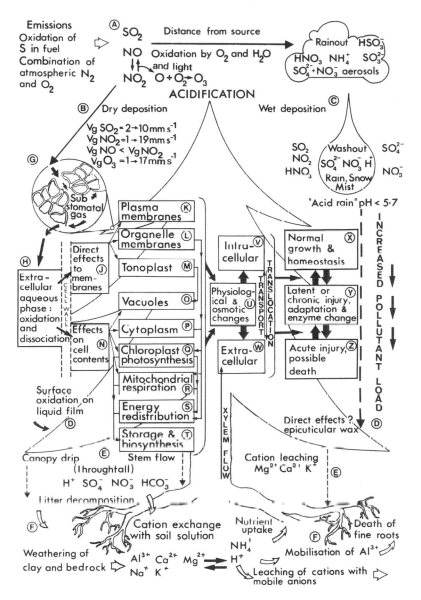

Fig. 7.6 Overview diagram showing the relationship between atmospheric pollution and vegetation. Note: see the accompanying text for a full description. (Courtesy also of Dr. P. H. Freer-Smith, UK Forestry Commission and CRC Press Inc., Florida.)

collectively known as the 'mesophyll resistance' (H). Although effects of pollutants such as ozone upon membranes (J) are understood rather better than those of other gases and their products, ultrastructural studies have indicated that acidic gases and their products have a range of effects on the plasma membranes (K), the tonoplast (M) and on the inner membranes (L), especially those of the chloroplast. Individual effects of sulphur dioxide and oxides of nitrogen within membranes have not been specifically identified but mixtures of sulphite and nitrite are highly conducive to enhanced free-radical formation and consequential leakiness similar to that caused by ozone alone.

Effects of air pollutants in solution have been seen upon cell constituents (N), as well as in membranes and, particularly, in the cytoplasm, (P), plastids (Q) and mitochondria (R). However, exchanges between the cytoplasm (P) and the vacuoles (O) across the tonoplast (M) are also important during detoxification and readjustments to buffering capacity. Imposition of change upon the intracellular pH of plant cells is one of the principal effects of the products of air pollutants in solution and has far-reaching implications upon metabolic controls mediated by means of cytoplasmic pH or proton gradients and fluxes. Observed biochemical effects upon various photosynthetic processes associated with plastids (Q) are both numerous and relatively well documented. There are a number of biochemical observations which indicate the likelihood of more-than-additive inhibitory effects of air pollutant mixtures upon growth. Effects upon respiration and mitochondrial processes (R) have been reported but these are not as great as those on photosynthesis. Thereafter, both types of organelle contribute to the bioenergetic status of the cell and to the availability of energy (S) expended either in storage, metabolism or biosynthesis (T).

The accumulation of effects upon membranes (J–M) and upon cellular and organellular processes (N–T) are reflected in an overall physiological change (U) which includes changes of photosynthesis, photorespiration and respiration as well as in osmotic, pH and nutrient status. Intracellular physiological events (V) will be revealed and altered by changes in extracellular events (W) associated with transport and translocation as well as by nutrient uptake and xylem flow. It would appear, especially in the case of pollution by oxides of nitrogen, that there are considerable influences of air pollutants upon physiological events, especially those of normal uptake and assimilation of nitrogen.

The overall success of the biochemical (K–T) readjustments and the physiological controls that stem from them (U–W), is mirrored by the final disposition of the whole plant and the various cellular components. With unaffected plants, or those only slightly affected by pollution, normally this would cause homeostatic readjustments (X) leading to natural growth, reproduction and differentiation of the plant within the environment. Further pollution would cause latent or chronic damage (Y) or so-called 'invisible injury' which is a consequence of diverted energy from growth to repair. The effectiveness by which adaptation of pollution sensitivity or tolerance is accomplished is measured by the balance between processes (X) and (Y). Failure to achieve and maintain a balance leads to acute damage or 'visible injury' and possible death of the plant (Z).

(c) Plants versus animals

At the cellular, biochemical or molecular level it is most noticeable that there are great similarities of response by both plants and animals to atmospheric pollution if one sets aside the overlying basic physiological differences between animals and plants. Passive gaseous uptake and the fluid circulatory systems in plants differ distinctly from active ventilation of the lungs (or gill movements) and circulation of the blood in animals. Similarly, photosynthetic sensitivity to free radicals of plants as against the strong binding to the haem of haemoglobin by carbon monoxide or hydrogen sulphide are clear differences associated with form and function. Apart from these special features, the mechanisms of entry of pollutants into living cells and a wide range of the biochemical reactions thereafter are, by and large, very similar especially those which mitigate or detoxify harmful free radicals. Indeed, the study of the effects of air pollutants on plants has many lessons and implications for those engaged in animal studies and *vice versa*, although this is rarely recognised by the two different groups of researchers.

Basic differences between animals and plants in the uptake and evolution of oxygen and carbon dioxide are fundamental to life and, to a certain extent, overshadow all else when gaseous exchanges are considered. Both plants and animals have a range of compounds containing haem groups which, in most cases, are highly protected. However, haemoglobin in the blood of animals binds carbon monoxide and hydrogen sulphide more avidly than oxygen. Certain plants also have leghaemoglobin in their nitrogen-fixing root nodules, which is capable of binding oxygen more strongly than haemoglobin, but it is most likely the fact that roots experience little carbon monoxide that causes animals to be more sensitive to these particular pollutants. However, in the case of sulphur dioxide, oxides of nitrogen and ozone, plants appear to be much more sensitive to these pollutants than animals. 'Why should this be so?' is a seemingly obvious question which is actually rarely posed or answered. The nearest one can get to an explanation is to compare two sets of calculations which contain a number of approximations.

For example, if 100 ppbv ozone-polluted air is taken as a starting point in each case, it is then possible to compare likely doses of a pollutant to both human lungs and plant mesophyll cells. At normal temperatures and pressure 22.4 litres of this polluted air would contain 2.24 μl (or 0.1 μmols) of pollutant. If the average human ventilation rate at rest is 480 litres h^{-1}, assuming tidal volume of 500 ml and 16 ventilations min^{-1} or 960 h^{-1} then the intake of pollutant to the lungs over an hour would be 2.14 μmol of ozone. However, under conditions of heavy exercise the breathing rate may increase eightfold (over the resting rate) and the expiratory reserve volume of 1.5 litres is used in addition to the tidal volume of 500 ml. This gives dramatic increases in ventilation rate (15 360 litres h^{-1}) and intake of pollutant (69 μmol h^{-1}). By taking the alveolar surface area to be 100 m^2 and the thickness of the cell layer between the alveolar space and the blood to be 0.5μm then the volume of this layer would be 50 ml and the dosage to it therefore ranges from 2.18 μmol per 50 ml per hour (or 43 μmol l^{-1} h^{-1}) at rest to 1.38 mmol l^{-1} h^{-1} during strenuous exercise.

In the case of ozone, uptake into plants largely depends upon stomatal access and the internal area of mesophyll cells available for uptake is between five and thirty times greater than the leaf surface area. Assuming the boundary layer has been stripped away by sufficient wind (i.e. greater than 4.8 km h^{-1}) then, for an alfalfa plant with a leaf surface of 14 m^2, the uptake rate through the stomata of 100 ppbv ozone-polluted air travelling with a speed of 4.8 km h^{-1} has been calculated to be 3.8 mmol h^{-1}. If the exposed internal cell surface area is taken as twenty-five times that of the leaf surface area and the thickness of the affected area is again 0.5 μm (i.e. a volume of 175 ml), it follows that the dosage to this sensitive area would be 3.8 mmol per 175 ml per hour or 21.7 mmol l^{-1} h^{-1}. Thus plants, despite passive uptake mechanism limited by diffusion, have their most sensitive tissues (i.e. those containing chloroplasts) exposed to more than sixteen times as much pollutant as those alveolar tissues of humans engaged in strenuous exercise even though the latter have active ventilation mechanisms. Differences of this magnitude (even allowing for the approximations) must be a major factor why plants are often more sensitive (relatively) than animals to atmospheric pollutants.

(d) Humans, fishes and other unfortunates

One measure of the success of a paper model is that it must be flexible and, with the minimum of modification, capable of adapting to cover

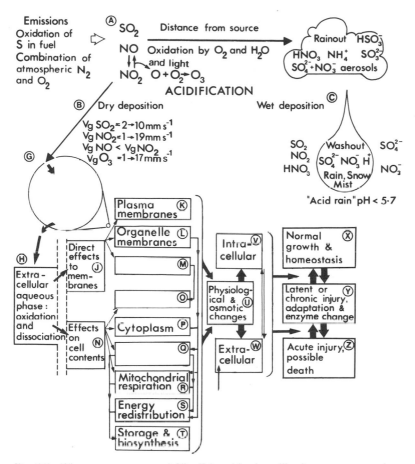

Fig. 7.7 The common core of Fig. 7.6, with plant-like features removed, before conversion into Fig. 7.8.

fresh situations. In the absence of a conceptual overview of changes in animals (as a result of atmospheric pollution) elsewhere in the literature, it is possible to take the plant paper model (Fig. 7.6), delete all the plant specific features (i.e. the backdrop of a tree plus roots, photosynthesis, stomata, etc.) to leave a common core (Fig. 7.7) and then to insert animal-like features such as lungs, fish gills, immune and circulatory systems. The outcome (Fig. 7.8) is highly satisfactory. The majority of the framework behind the different elements still applies. Fish, and their gill uptake mechanisms, show greatest responses to acidic wet deposition (C) into run-off waters (D) and to interference by aluminium ions (E). On the other hand, the active

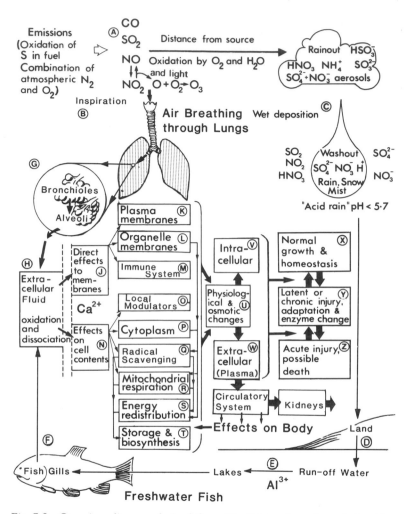

Fig. 7.8 Overview diagram derived from Fig. 7.6 to show the relationships between atmospheric pollution and animals. Once again the text covers the individual details.

inspiration (B) and expiration phases of air breathing are more affected by dry-deposition mechanisms. Uptake into the alveolar cells from the alveolar spaces (G) is through a extra-cellular liquid film (H) and, once again, the effects may be either on the membranes (J) or the cell contents (N). Free-radical scavenging (Q) is evident in both animal and plant tissues but effects on local modulators (M) may be especially important in any adjustments to physiological and osmotic changes (U) which are transmitted to the blood plasma (W). This has been shown to be especially important when effects of oxides of nitrogen are considered (Sect. 3.3d). Similarly, the effectiveness of immune responses (M), particularly those concerned with maintaining lung sterility, are known to be reduced in response to atmospheric pollutants such as sulphur dioxide (Sect. 2.3c). The active circulation of the blood permits effects of pollutants upon the alveolar cells to be modulated by the activities of other body tissues. This may mean detoxification and elimination by the kidneys or, for example, reduced calcification of developing bone tissues in fish fry (Sect. 4.4b). Notwithstanding such details, the essence of Fig. 7.8 remains the same as in Fig. 7.6. Any interference to the physiological and osmotic changes (U) caused by the action of the pollutant products on processes (H) to (T), either inside (V), or, outside (W) the alveolar (or gill) cells may then cause normal growth and homeostasis (X) to be transferred into chronic (Y) and acute (Z) injury.

(e)   Where next?

The obvious answer to this question, which is all too evident from the omissions of this text, is more research. But this would require experienced manpower, money and time. All are in short supply, particularly the last item because the answers for legislators are required now and not when it would be too late. Nevertheless, education and a wider appreciation of the problems are invaluable and it is to be hoped that this text will help to achieve this end. As stated in the Preface, references have been kept out of the text to assist clarity and selected bibliographies have been placed at the end of each chapter. A general bibliography appears after the Appendix and contains a list of the more general texts that could be consulted. If the question: 'Which texts are the most useful and which ones would one acquire or borrow?' is asked then those marked with an asterisk (*) provide a useful balance. When taken together, they develop a range of ideas in different directions.

SELECTED BIBLIOGRAPHY

Bauer, F. (ed.), *Diagnosis and Classification of New Types of Damage Affecting Forests,* (special edition). Commission of the European Communities, DGVI, F3, 1984, Brussels.
Bishop, J. A. and Cook, L. M. (eds.), *Genetic Consequences of Man Made Change.* Academic Press, 1981, London and New York.

Kettlewell, H. B. D., *The Evolution of Melanism.* Oxford University Press, 1973, London.

Lefohn, A. and Ormrod, D. P. *A review and Assessment of the Effects of Pollutant Mixtures on Vegetation:* Research recommendations. US. Environmental Protection Agency, EPA600/3–84–037, 1984, Corvallis, Oreg.

Levitt, J., *Responses of Plants to Environmental Stresses.* Academic Press, 1972, New York.

National Institute for Environmental Studies, *Studies on Effects of Air Pollutant Mixtures on Plants, Parts 1 and 2.* NIESJ nos. 65 and 66, 1984, Ibaraki, Japan.

Roberts, T. M., Darrall, N. M. and Lane, P., Effects of gaseous air pollutants on agriculture and forestry in the U.K. *Advances in Applied Biology* 9, 1983, 1–142.

Schulte-Hostede, S. (ed.), *Physiology and Biochemistry of Stressed Plants.* Proceedings of an International Workshop, GSF-Bericht 44/85, 1985, Munich.

Smith, W. H., *Air Pollution and Forests. Interactions between Air Contaminants and Forest Ecosystems.* Springer-Verlag, 1981, New York, Heidelberg and Berlin.

Ulrich, B. and Pankrath, J. (eds.), *Effects of Accumulation of Air Pollutants in Forest Ecosystems.* D. Reidel Pub. Co., 1983, Dordrecht.

Appendix

'Forget six counties overhung with smoke,
Forget the snorting steam and piston stroke.'

From the prologue to 'The Wanderers' by William Morris (1834–1896)

'Away with Systems!
Away with a corrupt world!
Let us breathe the air of the Enchanted Island.'

From 'The Ordeal of Richard Feverel' by George Meredith (1828–1909)

8.1 AIR POLLUTANT CONVERSIONS

The following BASIC program may be useful to those who wish to convert concentrations of a number of pollutants (HF, NH_3, NO, NO_2, O_3, PAN, H_2S, CO_2, CO and SO_2) from ppmv or ppbv into $\mu g\,m^{-3}$ or vice versa. The program allows for the correction of temperatures and pressures in a variety of different units. Furthermore, gaseous to aqueous conversions applicable to sulphur dioxide and nitrogen dioxide have been included. Other gaseous to aqueous conversions may be added if the appropriate factor is calculated and inserted on line 1430 and appropriate adjustments made to lines 1080, 1220 and 1280 allowing for rearrangement of the data order in lines 1410 and 1420. The program is written in Applesoft but only minor changes (e.g. to line 1100) are required to allow the use of this program on any microcomputer which operates BASIC. For those who just want a quick and rough guide to conversion and are not too concerned with temperature and pressures then Tables 8.1 and 8.2 (p. 242–3) may help.

```
1000   REM   ***************************************
1010   REM   * AIR POLLUTANT CONVERSIONS *
1020   REM   *      WRITTEN BY A.R.WELLBURN   *
1030   REM   * IF PROG.STOPS PRESS ANY KEY *
1040   REM   ***************************************
1050   DIM A$(12), A(12)
1060   FOR I = 1 TO 10: READ A$(I): NEXT I
1070   FOR I = 1 TO 10: READ A(I): NEXT I
1080   FOR I = 1 TO 2: READ B(I): NEXT I: RESTORE
1090   B$ = "*************************":C$ = "AIR POLLUTANT
CONVERSIONS"
1100   HOME : VTAB (3): HTAB (8): PRINT LEFT$ (B$, LEN (C$)):
PRINT : HTAB (8): INVERSE : PRINT C$: NORMAL : PRINT :
HTAB (8): PRINT LEFT$ (B$, LEN (C$))
1100   PRINT : PRINT: INPUT " Temperature (Degrees Celsius) ? ";T
1120   PRINT : INPUT : "Atmospheric pressure in Atmospheres (1), kPa
(2) or mmHg (3) ? ";E
1130   IF E < 1 or E > 3 THEN 1120
1140   IF E = 1 THEN F = 1
1150   IF E = 2 THEN F = 101.3
1160   If E = 3 THEN F = 760
1170   PRINT : INPUT "What is the pressure in the chosen units ? ";K
1180   PRINT : INPUT "Range required – PPMV (1) or VPB (2) ? ";A
1190   IF A + 1 THEN A$ = " PPM " :G = 1: GOTO 1210
1200   A$ = " PPBV":G = 1000: IF A < 1 or A > 2 THEN GOTO 1180
1210   PRINT : PRINT "Do you want to convert";A$;" to micrograms per
cubic meter (1) or the reverse (2) ": INPUT B
1220   IF B < 1 or B > 2 THEN GOTO 1210
1230   PRINT : FOR 1 = 1 to 10
1240   PRINT : PRINT A$(I); SPC( 25 - LEN (A$(I)));I
1250   NEXT I
1260   PRINT : PRINT : INPUT "Which pollutant (1–10) ? ";C
1270   PRINT : PRINT A$(C): PRINT LEFT$ (B$, LEN (A$(C)))
1280   IF B = 2 THEN GOTO 1340
1290   PRINT : PRINT "Amount of pollutant in";A$: INPUT P
1300   Q = ((P * A(C) / G) / .0224) * (273.15 / (273.15 + T)) * (K /
F):Q1 = ( INT (Q * 1000 +. 5)) / 1000
1310   PRINT : PRINT "Is equivalent to ";Q1;" micrograms per cubic
meter or ";Q1 / A(C);" nanomoles per 1"
1320   GET T$: IF C < O OR C > 2 THEN GOTO 1100
1330   GOTO 1380
1340   PRINT : INPUT "Amount of pollutant in micrograms per cubic
meter ? ";Q
```

1350 P = ((Q * .0224 * G) / A(C)) * ((273.15 + T) / 273.15) * (F / K)
: P1 = (INT(P*1000 +.5))/1000
1360 PRINT : PRINT "Is equivalent to ";P1;A$
1370 GET T$: IF C < 1 or C > 2 THEN GOTO 1100
1380 R = (INT (B(C) * SQR ((P * G)) * 1000 + .5) / 1000)):S = (INT
((R / A(C)) * 10000 + .5) / 10000)
1390 PRINT "or alternatively ";R;" PPMV or   ";S;"
millimoles/litre in solution."
1400 GET T$: GOTO 1100
1410 DATA   "Sulphur dioxide", "Nitrogen dioxide", Nitric
oxide","Ammonia","Ozone","PAN","Hydrogen fluoride","Hydrogen
sulphide", "Carbon dioxide","Carbon monoxide"
1420   DATA 64.066,46.008,30.008,17.029,48,105.051,20.008,34.082,44.01,
28.01
1430   DATA   10.3,358.9

8.2   FREE RADICALS AND SINGLETS

The term 'free radical' or simply 'radical' is used to describe those chemical species capable of independent existence which have one or more unpaired electrons in their outer electronic orbitals. Chemical bonds are formed when electrons 'pair up' and this mainly explains why free radicals are highly reactive. Some radicals may also carry an electrical charge and, as a consequence, their properties are also influenced by charge. However, there are several charged and neutral free radicals involved in air chemistry and atmospheric pollution. One example of a charged free radical is the superoxide free radical ($O_2^{\cdot-}$); the negative superscript signifying the charge and the dot indicating that it is a free radical because in one of two π^* antibonding orbitals around the superoxide molecule it has an unpaired electron as well as two paired electrons in the other π^* orbital. If one more electron is added to the outer orbital, to form two pairs of π^* electrons, then the free-radical nature is lost and the molecule becomes peroxide (O_2^{2-}).

The most stable oxygen molecule (called ground-state oxygen) is unusual in that it has one electron in each of the π^* antibonding orbitals, each having similar (or parallel) spin. By definition, ground-state O_2 (strictly $^3\Sigma\ O_2$) is thus a free radical in its own right and actually more reactive than superoxide. Even more reactive forms of the oxygen molecule may also exist because their spin restriction has been removed. These are often called singlet oxygen and may or may not be free radicals. One type ($^1\Delta g\ O_2$) has a pair of antiparallel

Table 8.1 Amounts of different pollutants in $\mu g\, m^{-3}$ equivalent to 1 vpm (1000 ppbv) at different temperatures but standard atmospheric pressure

Temperature (°C)	SO_2	NO_2	NO	NH_3	O_3	CO_2	CO	H_2S	HF
−5	2914	2092	1365	774	2183	2001	1274	1550	910
0	2860	2054	1340	760	2143	1965	1250	1521	893
5	2809	2017	1316	747	2104	1929	1228	1494	877
10	2759	1981	1292	733	2067	1895	1206	1468	862
15	2711	1947	1270	721	2031	1862	1185	1442	847
20	2665	1914	1248	708	1997	1831	1165	1417	832
25	2620	1882	1227	696	1963	1800	1146	1394	818
30	2577	1850	1207	685	1931	1770	1127	1371	805
35	2535	1821	1187	674	1899	1742	1108	1348	792

Table 8.2 Amounts of different pollutants in ppb; equivalent to 1000 $\mu g\ m^{-3}$ at different temperatures but standard atmospheric pressure

Temperature (°C)	SO_2	NO_2	NO	NH_3	O_3	CO_2	CO	H_2S	HF
−5	343	478	733	1291	458	500	785	645	1099
0	349	487	746	1315	467	509	800	657	1120
5	356	496	760	1339	475	518	814	669	1140
10	362	505	774	1364	484	528	829	681	1161
15	368	514	787	1388	492	537	844	693	1181
20	375	523	801	1412	501	546	858	705	1202
25	382	531	815	1436	509	556	873	718	1222
30	388	540	828	1460	518	565	888	730	1243
35	394	549	842	1484	527	574	902	742	1263

electrons in the π^* antibonding orbitals and is not a free radical although it is more reactive than ground-state oxygen, superoxide or peroxide. The most reactive oxygen molecule of all is the $^1\Sigma g^+$ singlet form which has an unpaired electron in each of the π^* antibonding orbitals, each having opposite spins to the other. The $^1\Sigma g^+$ form is so highly reactive that it decays immediately to the $^1\Delta g$ state so that only the latter is relevant to biological systems. Singlet oxygen is most readily formed during reaction of certain molecules with light (e.g. riboflavin, FAD, chlorophyll) which then allows the transfer of this photochemical energy to a nearby oxygen molecule. The most frequent biological reaction with singlet oxygen is attack upon carbon–carbon double bonds to form peroxides or dioxetanes. The latter are highly unstable and usually decompose to carboxyl groups.

As discussed in Ch. 5 (Fig. 5.7), oxygen may be reduced in a number of different ways. A one-electron reduction gives superoxide (Reaction 8.1) and two electrons form peroxide (Reaction 8.2). A four-electron reduction gives water (Reaction 8.3) but the splitting of hydrogen peroxide (the protonated form of peroxide), in light (Reaction 8.4) or in the presence of iron or copper (Reaction 8.5), forms one of the most reactive species possible – the hydroxyl radical $(OH^{\bullet})$.

Monoatomic oxygen, like some other atomic forms of gases (e.g. hydrogen, fluorine, nitrogen atoms), is actually also a free radical although the following dot superscipt is rarely used to indicate the presence of unpaired electrons – it is merely assumed. It too is capable of reacting with water to form hydroxyl radicals (Reaction 8.6).

$$O_2 + e \longrightarrow O_2^{\bullet -} \tag{8.1}$$

$$O_2 + 2e \longrightarrow O_2^{2-} \tag{8.2}$$

$$O_2 + 4H^+ + 4e \longrightarrow 2H_2O \tag{8.3}$$

$$H_2O_2 + light \longrightarrow 2OH^{\bullet} \tag{8.4}$$

$$Cu^+ \text{ or } Fe^{2+} + H_2O_2 \longrightarrow Cu^{2+} \text{ or } Fe^{3+} + OH^{\bullet} + OH^- \tag{8.5}$$

$$O + H_2O \longrightarrow 2OH^{\bullet} \tag{8.6}$$

Once formed, a free radical reacts with other components (e.g. Reaction 8.6 involving water) to form other free radicals, which may be more or less reactive than the original radical. Other examples include further reaction of hydroxyl radicals with hydrogen peroxide (Reaction 8.7) to form less-reactive superoxide radicals or reaction with anions, for example, like fluoride (Reaction 8.8) which would form fluorine atoms or even more reactive fluoride radicals (Reaction 8.9) or with any other component, organic or inorganic and inside or outside biological systems – so great is the reactivity of the hydroxyl radical.

$$OH^{\bullet} + H_2O_2 \longrightarrow H_2O + H^+ + O_2^{\bullet-} \qquad (8.7)$$

$$F^- + OH^{\bullet} \longrightarrow F^{\bullet} + OH^- \qquad (8.8)$$

$$F^{\bullet} + F^- \longrightarrow F_2^{\bullet-} \qquad (8.9)$$

8.3 PHOTOSYNTHESIS

One of the best ways to visualise the basic reactions of photosynthesis is to use a diagram (Fig. 8.1) which contains the elements of Van Niel's hypothesis of the 1930s updated to take account of modern developments. When excited by light, chlorophylls within the chloroplast membranes promote a charge separation (i.e. a polarisation) which remains distinct long enough for stable reductant to form by the conversion of oxidised nicotinamide dinucleotide phosphate (NADP$^+$) into reduced form (NADPH). In higher plants, this process is linked to water splitting or photolysis (Reaction 8.10) so that the evolved oxygen is effectively the product of the stable oxidant. In addition to the creation of NADPH, the stable reductant is responsible for trapping light energy into another form of chemical energy by converting inorganic orthophosphate and adenosine diphosphate (ADP) into adenosine triphosphate (ATP) by photophosphorylation (Reaction 8.11). Finally, as shown in Fig. 8.1, ATP and NADPH are mainly consumed by driving some of the reactions associated with the fixation of carbon dioxide into carbohydrate.

$$2H_2O + light \longrightarrow 4H^+ + O_2 + 4e \qquad (8.10)$$

$$ADP + P_i + light \longrightarrow ATP + H_2O \qquad (8.11)$$

One of the great virtues of Fig. 8.1 is that not only does it encompass all the main features of photosynthesis but that it also emphasises the dimensions of time. Moving from the left, events faster than 10^{-13} of a second are slowed down progressively to carboxylation reactions taking about one-tenth of a second (right-hand side). This is a most important feature of the photosynthetic apparatus in that it is organised to slow down fast reactions so that slower reactions which depend upon them can take place. The success with which it does this can be appreciated when it is realised that at least twelve orders of magnitude of time are involved.

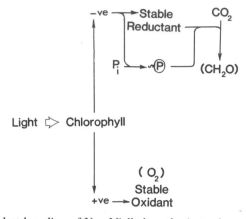

Fig. 8.1 Updated outline of Van Niel's hypothesis to show the basic events taking place in photosynthesis. Light falling on chlorophyll a promotes a charge separation which is stabilised and this enables the formation of reducing power and phosphorylation capacity which ensures the fixation of carbon dioxide into carbohydrate.

An additional consideration must be one of the relationship of photosynthesis to its location within plant tissues. In eukaryotes (i.e. organisms which have compartmentalised cells), such as higher plants, photosynthesis is confined to the chloroplasts. A diagrammatic representation of the major structural features of a chloroplast is shown in Fig. 8.2. The important points to note are that there are two major spaces – the stromal space and the lumen space (plus one minor one, the interenvelope space) separated from each other by a membrane. This is a most important feature which will be referred to later. Chlorophyll and other pigments (carotenes and xanthophylls) are concen-

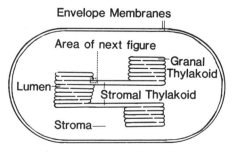

Fig. 8.2 Ultrastructural arrangement of the main components of a chloroplast. Two separate compartments exist (apart from the interenvelope space): the stroma and the lumen of the thylakoids.

trated within the internal membranes as complexes with proteins which are known as 'photosystems'. Individually, these membrane stacks are flattened balloon-like entities with connections one to another so that their lumens are interconnected. One classical wit has likened them to the baggy, pantaloon trousers or *thylakoides* of the Greek male national dress lying outside on grass to dry after washing – hence the use of the term 'thylakoid' for one element. Stacks of thylakoids form grana or granum stacks with stromal thylakoids (sometimes called 'fret membranes' or 'stromal lamellae') bridging across from granum to granum. The surrounding stroma contains a variety of enzymes; the most important of which are those involved with the fixation of carbon dioxide. These operate in a cyclical manner hence the term 'Calvin cycle' after its major discoverer. Other enzymes involved with starch or protein synthesis are also located in this region.

An important feature of early experiments with chloroplasts was the discovery of the Hill Reaction (Reaction 8.12) which demonstrated that photolysis and photoreduction of an electron acceptor, A, occurred together but were separable from fixation of carbon dioxide since early preparations of chloroplasts could carry out the Hill Reaction but could not fix carbon dioxide because they lacked stroma-retaining envelopes.

$$2H_2O + 4A \xrightarrow{\text{light}} O_2 + 4AH \tag{8.12}$$

Two photosystems consisting of chlorophylls associated with proteins are embedded within the thylakoid membranes (Fig. 8.3) and attached to the photosystems are other pigment complexes called light-harvesting chlorophyll–protein complexes or LHCPs. These light funnels increase the efficiency of light absorption and focus the energy upon special reaction centres within the photosystems.

In the sequence of reactions of photosynthesis, Photosystem II is before Photosystem I. This is always confusing but the reason is historical. In 1954, Arnon discovered cyclic photophosphorylation (Reaction 8.11), a process that used a chloroplast membrane component called Photosystem I. Three years later, he discovered non-cyclic photophosphorylation (Reaction 8.13) using a different photosystem which then had to be numbered Photosystem II.

$$4NADP^+ + 2H_2O + 2P_i + 2ADP \xrightarrow{\text{light}} 2ATP + 4NADPH + O_2 \tag{8.13}$$

Comparison of Reaction 8.13 with Reactions 8.10 to 8.12 will show that the essential elements of photolysis, photophosphorylation and the Hill Reaction are all encompassed within Reaction 8.13 – the $NADP^+$ acting as the acceptor, A, of Reaction 8.12 as it becomes photoreduced to NADPH. As a consequence, photolysis and Photosystem II are closely connected and the electrons flow from Photosystem II through a series of electron carriers to Photosystem I (where they are assisted on their way by a second light absorption), and finally arrive to reduce $NADP^+$ on the outer (i.e. stroma) side of the thylakoid membrane (Fig. 8.3). The exact chemical nature of the different electron carriers interposed between the photosystem is not pertinent to this level of description of photosynthesis but the action of one of them, plastoquinone, PQ, is worthy of explanation because it demonstrates an important event prior to ATP formation. In the reversible Reaction 8.14, the quinone part of plastoquinone picks up the electrons from Photosystem II (and also protons from the stroma) to form a reduced quinone or quinol (plastoquinol, PQH_2) before releasing the electrons back into the main flow towards Photosystem I. As it does so it releases the protons, this time into the thylakoid lumen (see Fig. 8.3), thereby re-forming oxidised plastoquinone once again.

$$PQ + 2e + 2H^+ \rightleftharpoons PQH_2 \qquad (8.14)$$

This brings us to one of the basic underlying features of bioenergetic processes which form ATP, such as photophosphorylation in chloroplasts and respiration in mitochondria. Indeed it is very useful to comprehend the principles involved because it unifies the two processes and leads to a better understanding of each of them. Both processes are linked by the pumping of protons vectorially across a membrane, in this case from stroma to lumen across the thylakoid, as a consequence of the electron flow within the membrane. The plastoquinone/plastoquinol couple clearly is part of this proton pumping but both photolysis on the lumen side producing protons and the photoreduction of $NADP^+$ which consumes protons on the stromal side obviously do the same thing. As a result, protons are added to the lumen making it more acidic and protons are taken away from the stroma making it more alkaline. In other words, a proton gradient or a pH difference is established across the thylakoid membrane. Hence the need for two distinct compartments – the stromal space and the lumen space – as mentioned earlier. This pH gradient may be considered to be available energy or, put another way, is capable of doing work. In one explanation of the consequences of this chemiosmotic mechanism, the pH gradient is subsequently harnessed by the protons flowing back through the coupling factor complexes (CF_1 and CF_0 on Fig. 8.3) to equalise the pHs and at the same time forcing the CF_1 particles to carry out photophosphorylation by forming ATP from ADP and phosphate. In order to

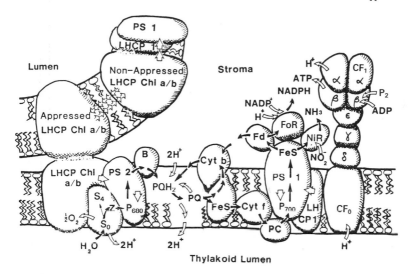

Fig. 8.3 Diagrammatic representation of the arrangement of the electron transport components within the chloroplast membranes (see Fig. 8.2 for cross-relationship). Flow of electrons is shown by solid arrows and proton or anionic movements are shown by open arrows. Light excitation in both photosystems (PS1 and PS2) is indicated by large open arrows and the sun–shade adaptive movements of large subunits (e.g. LHCP Chl a/b and LHCP, light-harvesting chlorophyll protein complex) by a sequence of arrows (upper left). Individual electron flow components are marked with their usual abbreviations: S_0, S_4 and Z are important components of the water-splitting and oxygen-evolution mechanisms; B is a polypeptide sensitive to herbicides; PQ, plastoquinone; cyt, cytochromes *b* and *f*; FeS, an iron-sulphur centre similar to Fd or ferredoxin; PC, plastocyanin, a copper protein; NiR, nitrite reductase and FoR, ferredoxin oxido-reductase. The appropriate subunits of the coupling factor, CF_1, protruding into the stroma from the CF_0 units are responsible for ATP formation. (Courtesy of Academic Press, Florida).

distinguish electron flow from proton flow, solid arrows have been used for electron flow and open arrows for proton flow in Fig. 8.3. The essential point to note is that, although electron flow is responsible for the generation of proton flow, the two flows are at right-angles to each other and ATP formation is only dependent upon proton flow. Conceptually, we may summarise this in Fig. 8.4 which clearly forms a major part of the ideas originally encapsulated in Fig. 8.1.

The remaining phase of photosynthesis is fixation of the carbon dioxide, the generation of sugars and the storage of these products in the form of starch. At this level of description it would not be

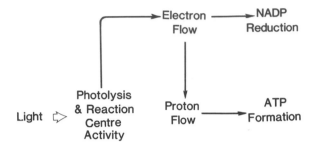

Fig. 8.4 Simplified outline of the basic elements within Fig. 8.3 which may be related directly to the modified Van Niel hypothesis diagram (Fig. 8.1).

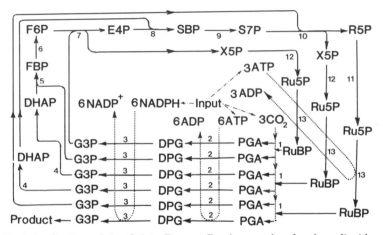

Fig. 8.5 Outline of the Calvin-Benson-Bassham cycle of carbon dioxide fixation to illustrate the fate of carbon atoms. Note especially that the inputs (centrally placed) consist of reduced NADPH, ATP and carbon dioxide and, although most fixed carbon is recycled back to the starting point, the excess is released as product (lower left). The abbreviations used are as follows: RuBP, ribulose-1,5-*bis*-phosphate; PGA, 3-phosphoglycerate; DPG, 1,3-diphosphoglycerate; G3P, glyceraldehyde-3-phosphate; DHAP, dihydroxy-acetone phosphate; FBP, fructose-1,6-*bis*-phosphate; F6P, fructose-6-phosphate; E4P, erythrose-4-phosphate; SBP, sedoheptulose-1,7-*bis*-phosphate; S7P, sedoheptulose-7-phosphate; X5P, xylulose-5-phosphate; R5P, ribose-5-phosphate; and Ru5P, ribulosephosphate. The numbered enzymes are: 1, RubisCO or ribulose-1,5-*bis*-phosphate carboxylaseoxygenase; 2, phosphoglycerokinase; 3, triosephosphate dehydrogenase; 4, phosphotriose isomerase; 5 and 8, aldolase; 6, fructose-*bis*-phosphatase; 7 and 10, *trans*-ketolase; 9, sedoheptulose-*bis*-phosphatase; 11, 12 and 13, phosphopentose isomerase, epimerase and kinase, respectively.

appropriate to cover in detail all the reactions of the Calvin cycle or the individual metabolites involved. The important features shown in Fig. 8.5 emphasise that, whilst the main process is cyclical, the surplus carbon is siphoned out mainly as phospho-3-glyceraldehyde and that NADPH and ATP are consumed to drive the cycle round at various stages and in unequal proportions ($3ATP : 2NADPH : 1CO_2$). More ATP is required for starch synthesis. This partially explains why both cyclic (Reaction 8.11) and non-cyclic (Reaction 8.13) photophosphorylations are required.

8.4   OXIDATIVE PHOSPHORYLATION AND RESPIRATION

Mitochondria are cylindrical organelles ($3-5\ \mu m$ long $\times\ 0.5-1\ \mu m$ diameter) with double membrane envelopes enclosing the matrix space (see Fig. 8.6). The inner envelope membrane may fold inwards to form cristae in a number of different ways. In animals, they are generally in the form of shelf-like structures. The space inside the shelves is called the 'intracristal space' and this is in continuum with the 'inter-envelope space'. Together these spaces are called the 'intermembrane space' which is quite distinct from the 'matrix space' and a vital

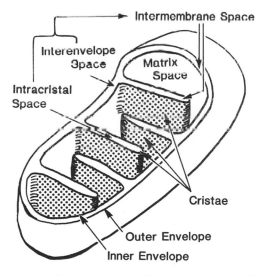

Fig. 8.6 Ultrastructural arrangements of a mitochondrion. The two major compartments are the inner matrix space and the intermembrane space.

feature in any chemiosmotic explanation of ATP formation in mito-chondria, just as in chloroplasts (see Sect. 8.3).

The matrix spaces within mitochondria are the location of enzymes of the Krebs' tricarboxylic acid cycle in which pyruvate is oxidised to carbon dioxide and water. This series of oxidations in the matrix space, along with the equally important 'beta-oxidation helix' of fatty acid degradation, form a number of reduced coenzymes which initiate a cascade of successive oxidations and reductions terminating in the reduction of molecular oxygen to water. This cascade of electron carriers is located within the cristae of the mitochondria and is known as the 'electron transport' or 'respiratory' chain (Fig. 8.7). At certain points on the same membrane, ATP is produced from ADP as an indirect consequence of the electron flow along the respiratory chain. This overall process is known as 'oxidative phosphorylation' in animals and plants to signify the terminal consumption of oxygen. In certain microbes, oxygen is not the terminal electron acceptor of respiratory electron flow so this process should be called more correctly 'respiratory chain phosphorylation' to allow for modes of respiration which use alternative acceptors such as sulphate, nitrate, nitrite, ferric iron, etc. (see Chs 2 and 3).

Various reactions in the Krebs' tricarboxylic acid cycle reduce the cofactors nicotinamide adenine dinucleotide (NAD^+/NADH) or flavin adenine dinucleotide (FAD/$FADH_2$) attached to different enzymes embedded in the cristae of mitochondria. Good examples of these (also shown in Fig. 8.7) are the following reactions:

$$\alpha\text{-ketoglutarate} + \text{Enz-NAD}^+ + \text{CoA} \qquad (8.15)$$
$$\longrightarrow \text{succinyl-CoA} + \text{Enz-NADH} + CO_2$$

$$\text{succinate} + \text{Enz}'\text{-FAD} \longrightarrow \text{fumarate} + \text{Enz}'\text{-FADH}_2 \quad (8.16)$$

The electron flow, induced by this type of reaction, cascades down a redox chain of electrochemical couples, from highly reduced components with electronegative potentials of high energy to oxidised compounds with electropositive potentials of much lower energy content, ending with the reduction of oxygen to form water. Some of the reactants carry both electrons and protons but others, such as the cytochromes, transport only electrons on the iron atoms of their haem constituents. An important electron transport intermediate is ubiquinone, UQ, which undergoes the following reduction:

$$UQ + 2e + 2H^+ \rightleftharpoons UQH_2 \qquad (8.17)$$

As shown in Fig. 8.7, ubiquinone picks up electrons from enzymes linked to FADH and protons from the matrix space to form ubiquinol, UQH_2. This reduced quinone then releases electrons onto the cyto-

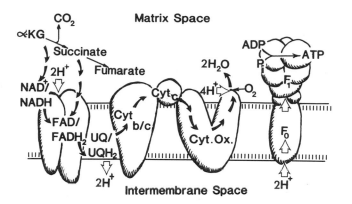

Fig. 8.7 The molecular organisation of the mitochondrial cristae membrane to illustrate electron flow (solid arrows) and movement of protons (large open arrows). Abbreviations are as follows: α-KG, α-ketoglutarate; UQ, ubiquinone; cyst$_{b/c}$, cyt$_c$ and cyt oxid(ase), different cytochromes; and F_0/F_1 are the coupling factors which use the proton gradient to form ATP.

chromes and protons into the intermembrane space. This feature is identical to one pointed out before (see Sect. 8.3) in connection to the chloroplastidic process of photophosphorylation involving plastoquinone. As electrons flow, protons are pumped vectorially across the inner mitochondria (or plastid) membranes making the intermembrane space of mitochondria more acidic and the matrix space more alkaline by a chemiosmotic mechanism.

The proton gradient (or pH difference) and a charge separation (or membrane potential) established across the membrane during electron flow combine together to give a proton motive force which is capable of doing work. This proton motive force may be used to change the conformation of the mitochondrial coupling factor particles (F_1 on Fig. 8.7) protruding from the cristae into the matrix space (which are attached to F_0 embedded in the membrane). These altered components are then able to convert ADP and orthophosphate into ATP and at the same time dissipate the proton motive force.

ATP inside the matrix space of the mitochondria is then made available to the rest of the cell by adenine nucleotide carrier proteins or translocators embedded in the envelopes which swap, in a one-for-one manner, ADP coming in from the cytoplasm for ATP going out from the matrix space. By such means oxidative phosphorylation in a mitochondrion process provides sufficient energy in the form of ATP to the rest of the cell.

8.5 ACIDITY, pH, pK_a AND MICROEQUIVALENTS

A substance that is able to donate protons (H^+ – more strictly hydroxonium ions, H_3O^+) to a solution is defined as an acid. Conversely, a base is a substance that accepts protons. These definitions are useful because they parallel those used to define reduction and oxidation (see Sect. 2.1c) and emphasise the close thermodynamic relationships that exist between the two types of phenomena. By convention, a substance is enclosed within square brackets if a concentration of that substance is also implied: thus $[H^+]$ indicates a concentration of protons in moles per litre. In Reaction 8.18, for example, the equilibrium constant ($K_{8.18}$) governing the ionisation of acetic acid is given by the following relationship:

$$CH_3COOH \rightleftharpoons CH_3COO^- + H^+ \qquad (8.18)$$

$$K_{8.18} = \frac{[CH_3COO^-]\,[H^+]}{[CH_3COOH]}$$

Thus if $[H^+]$ and the equilibrium constant are known (i.e. $K_{8.18} = 2.24 \times 10^{-5}$ M) it is possible to calculate the ratio between the charged and uncharged forms of acetic acid. Another way of looking at this is to appreciate that the concentration of the charged and uncharged forms is governed by $[H^+]$.

The usual way of expressing $[H^+]$ is on a logarithmic scale (known as the pH scale) because very large ranges of concentrations may exist where pH is defined as:

$$pH = \log \frac{1}{[H^+]} = -\log [H^+]$$

This gives a scale of pH values from 0 to 14 but biological systems usually involve pH changes within the range 2 to 9.

If, in the acetic acid example, the charged $[CH_3COO^-]$ and uncharged $[CH_3COOH]$ forms of the acid are equal to each other then it follows that:

$$[H^+] = K_{8.17}$$

The pH at which this occurs is known as the pK_a value and is the point of maximum sensitivity to pH change. In this case, as $K_{8.18}$ is equivalent to 2.24×10^{-5} M, then this particular pH or pK_a value is 4.65. It is also a useful indicator of the strength of an acid. The stronger the acid the more ready it is to donate protons and hence the more protons are required to reverse this action. A number of important ionisations relevant to air pollution and their pK_a values are shown in Table 8.3.

Table 8.3 pK_a values for acids in aqueous solution

System	Equilibrium			pK_a
Sulphuric acid	H_2SO_4	$\rightleftharpoons$	$H^+ + HSO_4^-$	< -2
Nitric acid	$HONO_2$	$\rightleftharpoons$	$H^+ + NO_3^-$	-1.4
Sulphurous acid	H_2SO_3	$\rightleftharpoons$	$H^+ + HSO_4^-$	1.8
Bisulphate	HSO_4^-	$\rightleftharpoons$	$H^+ + SO_4^{2-}$	2.0
Hydrofluoric acid	HF	$\rightleftharpoons$	$H^+ + F^-$	3.3
Nitrous acid	$HONO$	$\rightleftharpoons$	$H^+ + NO_2^-$	3.3
Aluminium	$AL(H_2O)_6^{3+}$	$\rightleftharpoons$	$H^+ + Al(H_2O)_5(OH)^{2+}$	5.0
Carbonic acid	$H_2SO + CO_2$	$\rightleftharpoons$	$H^+ + HCO_3^-$	6.4
Hydrogen sulphide	H_2S	$\rightleftharpoons$	$H^+ + HS^-$	7.1
Bisulphite	HSO_3^-	$\rightleftharpoons$	$H^+ + SO_3^{2-}$	7.2
Ammonium	NH_4^+	$\rightleftharpoons$	$H^+ + NH_3$	9.3
Bicarbonate	HCO_3^-	$\rightleftharpoons$	$H^+ + CO_3^2$	10.3
Bisulphide	HS^-	$\rightleftharpoons$	$H^+ + S^2$	12.9
Water	H_2O	$\rightleftharpoons$	$H^+ + OH^-$	14.0

Values of pK_a in relation to pH are useful as a quick guide because a move of 1 pH unit away from the pK_a value changes the ratio between the charged and uncharged forms of an acid by a factor of 10.

One way of calculating these changes more accurately over the whole range of pH values, especially if they do not differ by whole pH numbers, is to use the Henderson–Hasselbach equation which has a number of different forms summarised as follows:

$$pH = pK_a' + \log \frac{[salt]}{[acid]}$$

or:

$$pH = pK_a + \log \frac{[base]}{[salt]}$$

In the case of Reaction 8.18, if the pH is measured as 2.65, the following equation is relevant:

$$pH = pK_a + \log \frac{[CH_3COO^-]}{[CH_3COOH]}$$

and substituting gives:

$$2.65 = 4.65 + \log \frac{[CH_3COO^-]}{[CH_3COOH]}$$

or:
$$-2 = \log \frac{[CH_3COO^-]}{[CH_3COOH]}$$

therefore:

$$\text{antilog of } -2 = \frac{[CH_3COO^-]}{[CH_3COOH]} = 0.01$$

Hence the ratio of uncharged to charged acid is 1 : 100. On back-checking, this must be so because an increase of $[H^+]$ one hundred-fold over the value at the pK_a has taken place.

Ionisation of water (Reaction 8.19) is governed by the appropriate equilibrium constant, $K_{8.18}$. But because $[H_2O]$ is so large (55.5 molar or 55.5 M)[1] in relation to $[H^+]$ or $[OH^-]$ and declines imperceptibly, it is often treated as a constant. By multiplying $K_{8.18}$ by $[H_2O]$ one obtains the ionic product of water (K_w) as follows:

$$H_2O \rightleftharpoons |H^+ + OH^- \tag{8.19}$$

$$K_{8.18} = \frac{[H^+][OH^-]}{[H_2O]}$$

$$K_{8.18} \times [H_2O] = K_w$$

or:

$$K_w = [H^+][OH^-] = 10^{-14} M^2$$

When the $[H^+]$ equals that of $[OH^-]$, i.e. $[H^+]^2$, this will give a neutral solution. Thus substituting in the above:

$$[H^+]^2 = 10^{-14} M^2$$

therefore: $[H^+] = 10^{-7}$ M, or pH = 7

Thus at pH 0 $[OH^-]$ will be negligible and $[H^+]$ will be 1 M for simple equilibria like those shown in Reactions 8.18 and 8.19.

Another way expressing $[H^+]$ is in terms of equivalents because, in some cases, certain acidic reactions (e.g. Reaction 8.20) produce more protons than simpler reactions (e.g. Reactions 18 or 8.19) and

$$H_2SO_4 \rightleftharpoons 2H^+ + SO_4^{2-} \tag{8.20}$$

---

[1] Note: Conventionally, when equilibria involve H_2O as a reactant, $[H_2O]$ is actually set at 1 M and not 55.5 M (i.e. 1000 g l^{-1} divided by the molecular weight of 18.016) and all equilibrium constants involving water as a reactant therefore need to be adjusted by a factor of 55.56 molar if other factors such as rate constants, etc., are under consideration.

the identity of the acid or its products is sometimes unknown. The relationship between pH and microequivalents is given by:

$$pH = 6.0 - \log (\mu\text{equivalents } H^+ \, 1^{-1})$$

Thus at pH 0 there is one equivalent of protons per litre, at pH 3 one milliequivalent, and pH 6 one microequivalent.

8.6 HISTORICAL AND ETYMOLOGICAL POSTSCRIPT

The word 'gas' was invented by the Flemish chemist Van Helmont (d. 1644) taking as his inspiration the Dutch word *geest*, meaning spirit or ghost, which is related to the German word *geist*. He also used another word for gas, *blas*, from the Dutch verb *blazen* – to blow but this was never taken up. Both words invented by Van Helmont (who incidentally was also the first to describe the gas which was subsequently shown to be carbon dioxide) alluded to the Greek word, *chaos*, meaning empty space.

The word 'pollute' comes directly from the Latin word, *pollutus* – the past participle of *polluere*, meaning to defile derived from *por* or *port* that is towards (a river bank) and *luere* to wash over. The Latin word *pollutionem* was first used in 1157.

'Atmosphere' is a coined word from the Greek *atmos*, meaning breath or vapour, and *sphaira*, a sphere, whilst *air* comes from the Greek word *aer*, for haze or mist and through Latin *aer* (air) and French *air*. 'Acid' comes directly from the Latin *acidus*, meaning sour and 'rain' through Anglo-Saxon *regn*, similar to the Dutch and German words *regen*, which have been connected to the Latin *rigare* meaning to moisten.

'Oxygen' is a coined word dating from 1744 meaning generator of acids using the Greek word *oxus*, meaning sharp, pungent or acid, and *gene*, to produce. Discovery of oxygen is usually credited to the Swedish chemist Scheele in 1771 but the so-called 'Father of plant physiology', Stephen Hales had already detected its presence in 1727. 'Ozone' was the name derived in 1840 by the Swiss chemist Schönbein from the Greek *ozein*, to smell (from which the Latin *odor*, a smell, also arises) to describe the odour developed in an atmosphere during electrical discharge.

Credit to the one who discovered each of the different gases is a difficult and contentious subject because very often no one scientist did all the vital work. Very often it was achieved as a series of observations made by several scientists; quite often over a long period. But the name of Joseph Priestley (1733–1804) is without doubt one of the most important. In a relatively short period just over 200 years ago he achieved a most remarkable set of discoveries. In his studies he showed that a constant volume of inert gas remained after combustion whilst Rutherdorf elsewhere (and unbeknown to each other) showed the same by analysing gas exhaled during respiration. This inert gas they called 'mephitic' or 'phlogisticated air'. They are often credited as being the co-discoverers of nitrogen (a derived word which harks back through the Old French word for saltpetre, *nitre*, through the Latin *nitrum* and the Greek *nitron* – both meaning native soda – to the Egyptian *netra*, meaning to cleanse). It was left to the Swedish chemist Scheele to show in 1777 that air consisted principally of 20 % oxygen and 80 % nitrogen. Meanwhile, Priestley obtained the first samples of nitrous oxide, N_2O, in 1772 as well as ammonia in 1774 and the French chemist Lassonne found carbon monoxide in 1776. The word 'ammonia' is a contraction of *sal ammoniac* derived from the Latin *ammoniacum* and the Greek *ammoniakon*, meaning salt of Ammon, the God Jupiter who was originally the Egyptian Amen, the Hidden One.

The pungent gas now known as 'sulphur dioxide' has been known and used by man since the earliest records were kept. The English word 'sulphur' comes directly from Latin *sulphur*, and the American variant 'sulfur' from the Latin variant of *sulphur*, *sulfur*. Both are variants of the earlier Latin word *sulpur*. Each language is therefore unable to claim authenticity – both are misspellings. Sulphur dioxide is also involved in the first deliberate act of atmospheric pollution by man and the subject of the first record of a distinct biological effect. The blind Ionian Greek writer Homer, the traditional author of the *Iliad* and the *Odyssey*, was thought to have lived and died sometime between 1050 and 850 B.C. In his writings he refers to the deliberate burning of sulphur by the Egyptians, a practice that he claims had been undertaken previously for over a thousand years (i.e. before 2000 B.C.). They did this, he wrote, in order to generate a pungent gas which bleached their linen and killed the parasites residing there at the same time.

The first record of harmful and deliberate effects of a gas upon humans also comes from Greek sources. The Oracle at the Temple of Apollo at Delphi was known as the Pythoness or Pythian because of the legendary giant python slain on Mount Parnassus by Apollo. A trance-like or mesmeric state was required before the Pythian could predict coming events and this was induced by carbon dioxide emitted from fissures in the rock nearby. Not surprisingly, the Pythians had to be replaced frequently because of the dizziness, nausea, delirium or death that ensued. Plutarch (46–120 A.D.) tells of there being at least three Pythians on duty at any one time to supply the necessary reserves.

I hope you have enjoyed the quotations used throughout this book, some were not written originally with pollution in mind but appropriate nevertheless. One of the most outspoken was undoubtedly the following:

'The Aire[1] below is double dyed and damned;
the air above, with lurid smoke is crammed;
the one flows steaming foul as Charon's Styx[1]
its poisonous vapours in the other mix.'

Quoted by William Osburn at a meeting of The Leeds Philosophical and Literary Society in 1857.

[1] Rivers flowing through Leeds and nine times round the Underworld of Greek mythology, respectively).

General bibliography

*Brimblecombe, P., *Air Composition and Chemistry*. Cambridge University Press, 1986, Cambridge.

Brunner, C. R., *Hazardous Air Emissions from Incineration*. Chapman & Hall, 1985, New York.

Grefen, K. and Reinisch, D. W. and Suess, M. J. (eds.), *Ambient Air Pollutants from Industrial Sources: A Reference Handbook*. Elsevier Sci Pub., 1984, Amsterdam.

Guderian, R. (ed.) *Air Pollution by Photochemical Oxidants*. Springer-Verlag, 1985, New York.

*Halliwell, B. and Gutteridge, J. M. C., *Free Radicals in Biology and Medicine*. Clarendon Press, 1985, Oxford.

Hamilton, A. and Hardy, H. L., *Industrial Toxicology* (3rd ed). Publishing Sciences Group, 1973, New York.

Henderson-Sellers, B., *Pollution of our Atmosphere*. A. Hilger, 1984, Bristol.

Koziol, M. J. and Whatley, F. R. (eds.), *Gaseous Air Pollutants and Plant Metabolism*. Butterworths, 1984, London.

Lee, S. D. (ed.), *Biochemical Effects of Environmental Pollutants*. Ann Arbor Sci. Pub., 1977, Ann Arbor, Mich.

Legge, A. H. and Krupa, S. V. (eds.), *Air Pollutants and their Effects on the Terrestrial Ecosystem*. Wiley Interscience, 1986, New York.

Mansfield, T. A. (ed.), *Effects of Air Pollutants on Plants*. SEB Seminar Series no. 1. Cambridge University Press, 1976, Cambridge.

Smith, W. H., Air Pollution and Forests: Interactions between Air Contaminants and Forest Ecosystems. *Springer–Verlag, 1981, New York, Heidelberg and New York.

Stern, A. C., *Air Pollution* (3rd edn), vol. 2. *The Effects of Air Pollution.* Academic Press, 1977, New York.

Treshow, M. (ed.), Air Pollution and Plant Life. *John Wiley, 1984, Chichester.*

Turiel, I., *Indoor Air Quality and Human Health.* Stanford University Press, 1985, Palo Alto, San Francisco, Calif.

Unsworth M. H. and Ormrod, D. P. (eds.), *Effects of Gaseous Air Pollution in Agriculture and Horticulture.* Butterworths, 1982, London.

Waldbott, G. L., Health Effects of Environmental Pollution *(2nd edn). C. V. Mosby Co., 1978, St. Louis, Mo.*

---

[1] See Sect. 7.4(e) for explanation of use of asterisks.

Index